# THE IMMORTAL CELL

### WHY CANCER RESEARCH FAILS

## DR. GERALD B. DERMER

AVERY PUBLISHING GROUP INC.

Garden City Park, New York

Cover designers: Rudy Shur and Ann Vestal
Cover photo supply house: The Stock Market
Cover photograph © by Howard Sochurek
In-house editors: Barbara Conner and Amy C. Tecklenburg
Typesetter: Bonnie Freid
Printer: Paragon Press, Honesdale, PA

**Library of Congress Cataloging-in-Publication Data**

Dermer, Gerald B.
    The immortal cell : why cancer research fails /Gerald B. Dermer.
      p. cm.
    Includes bibliographical references and index.
    ISBN 0-89529-582-2
    1. Cancer—Research—United States.  2. Cell lines.  3. Cancer
cells.  I. Title.
RC 267.D47  1993
616.99'4'0072—dc20                     93-26852
                                     CIP

Printed in the United States of America.

10  9  8  7  6  5  4  3  2  1

# Contents

*For Emilie and Mireille*
*May their generation have the treatments*
*their grandmothers' did not.*

# Acknowledgments

I have been obsessed with this book for three years. The encouragement and sense of humor of my wife, Rosalind, have lightened my mood on the many days I was convinced my efforts would lead nowhere. Her suggestions have greatly improved the manuscript. Rosalind's commitment to the goals of this project is without equal. I would also like to thank my stepson, Brett Macune. He never complained on the many occasions I needed the computer and he had to vacate his room. My appreciation also goes to Dr. Howard Reuben, who reviewed an early draft of the manuscript and offered several suggestions. Catherine A. Heusel edited the book, and her deft touch is on every page. Rudy Shur, managing editor of Avery Publishing Group, has always been enthusiastic about the book. I am grateful that he believed the public needed to know this story.

# Preface

I have spent about one-half of my adult life as a scientist in cancer research. From the world of the surgical pathology lab—where I studied tumors removed from the bodies of living cancer patients—to the rarefied world of medical school research on cells living in petri dishes, I have immersed myself in knowledge about one of the greatest scourges of our time. And I have learned that there is a vast and deadly gap between the reality of cancer, which strikes human beings, and the theory of cancer, which thousands of researchers are using in their search for a cure.

There is an East Indian folktale, about a group of blind men and an elephant, that is tragically descriptive of the state of cancer research today. In the fable, several blind men are asked to describe an elephant. One, after feeling the elephant's trunk, asserts that it is "very like a snake." Another, feeling a leg, insists it is "very like a tree." Another, feeling the tail, insists it is "very like a rope." Each leaves the scene certain that his impression is correct.

In cancer research, most scientists are in a far worse position than the blind men of the fable. The blind men were gathering what data they could from an actual elephant. Most cancer researchers, using all their faculties, are investigating an unnatural "animal" created in the laboratory, mistakenly applying their data to the real "elephant" of human cancer.

Although some of my colleagues are aware of this gap, few are willing to risk their careers by discussing it openly. In the absence of public debate, cancer scientists around the country are free to propagate the myth of a productive "war on cancer." No one wants to admit that this so-called war has been a worthless investment of taxpayers' money and scientists' time. But as more and more money is spent, with fewer and fewer meaningful results, increasing numbers of patients and their families, taxpayers, and politicians want to know the reasons why.

The answers can be found within the hallowed halls of the National Cancer Institute and other bastions of the cancer establishment, where well-funded scientists are tilting at the molecular windmills of their favorite laboratory representation of cancer—cells growing in petri dishes. Almost everything in science and medicine, including the development of effective treatments, hangs on the reliability of these experimental models. In cancer, the use of these unnatural cells as a model for the human disease has been directly responsible for our ongoing defeat. The cultures, termed *cell lines*, give incorrect and clinically useless information about cancer.

I came to the world of academia from a hospital environment, after more than a decade of working with tumors removed from the bodies of living cancer patients. The discrepancies between what I knew of human cancer and what I read in research journals and saw in researchers' petri dishes were so striking that I could not keep silent. But I soon learned the fate of those who would challenge a fashionable and very productive research model, however incorrect it might be.

Although it is shielded from the public by high-minded pronouncements and scientific jargon, the cancer establishment is afflicted with a mental and moral malaise. It is more interested in maintaining the status quo than in finding the answers to the cancer riddle, and will defend that status quo against all comers. Its struggle to retain credibility and power may well last decades and cost millions of lives, unless the source of its funding—the taxpaying public—demands reform.

I tried to interest the media in the problems in the cancer industry. I wrote letters and made phone calls, but no one wanted to get involved. The medical reporter from the *Arizona Republic* was concerned that if he wrote stories critical of cancer research, the Cancer Center of the University of Arizona School of Medicine would no longer give him its stories. I encountered similar resistance throughout the media. Finally, in the summer of 1989, I realized that if this story were ever to be made known to the public, I would have to do it myself. I knew that the most effective force for change is informed citizens demanding it. The result is *The Immortal Cell*.

This book was written to alert the public to the truth behind the failed war on cancer. I wrote it as neither a journalist nor a practicing physician, but rather as a research scientist who has seen the truth about cancer research firsthand. In these pages I am harshly critical of much of our medical science establishment. I do not laud dedicated researchers, nor do I paint a rosy picture of soon-to-be-discovered cancer cures. Although I firmly believe that research can and will produce practical and effective treatments for cancer, such advances will never come from the present research paradigm. Instead, I present a story of narrow-mindedness, vaulting ambition, and self-interest among those to whom we have given our trust—cancer scientists. It is an account of a scientific and medical scandal of the highest order. But most of all, it is a tale of poor science and the pressures that induce cancer scientists to do unsound work.

*The greatest derangement of the mind is to believe in something because one wishes it be so.*

Louis Pasteur

*False facts are highly injurious to the progress of science, for they often endure long.*

Charles Darwin

# 1

# Losing Ground

*Why has the outcome of sixty years of work by many first rate scientists and at a cost of hundreds of millions of dollars had so insignificant an influence on the prevention or treatment of cancer?*[1]

Sir Frank Macfarlane Burnet
*Genes, Dreams and Realities,* 1971

*Why is cancer the only major cause of death for which age-adjusted mortality rates are still increasing?*[2]

John Bailar, M.D., Ph.D., and Elaine Smith, Ph.D.,
epidemiologists, 1986

*We express grave concerns over the failure of the "war against cancer" [and] the absence of any significant improvement in the treatment and cure of the majority of all cancers . . . the cancer establishment and major pharmaceutical companies have repeatedly made extravagant and unfounded claims for dramatic advances in the treatment and cure of cancer.*[3]

From a statement issued by sixty top authorities
in preventive medicine and public health
at a Washington, D.C., press conference
February 4, 1992

3

This year more than 1 million Americans will be diagnosed with cancer. More than half a million will die. Breast cancer alone will be detected in 175,000 women. Since 1950, the overall incidence rate for cancer has increased by 44 percent, and the American Cancer Society has estimated that more than 75 million of the people alive today will eventually develop some form of cancer.[4] At this rate, cancer will surpass heart disease as the leading cause of death before the turn of the century.

The reality of cancer, the pain and suffering of individual patients and their families, can never be measured by mere statistics. Those who agonize through poisonous, nauseating, and ultimately ineffective treatments can hardly be encapsulated in numbers. They are the casualties of a twenty-year war on cancer that has won few significant battles.

## A DECLARATION OF WAR

On August 5, 1937, the Seventy-Fifth Congress of the United States created the National Cancer Institute (NCI). As a division of the United States Public Health Service, the institute was, under the provisions of the National Cancer Act of 1937, to "conduct, assist and foster researches, investigations, experiments and studies relating to the cause, prevention, methods of diagnosis and treatment of cancer."[5] With public concern over the cancer problem on the rise, the government had decided it was time to take the aggressive lead in cancer research. Before 1937, financial aid for cancer research came from modest amounts of private philanthropy.

More than three decades later, in 1971, Nobel prize-winning immunologist Sir F. Macfarlane Burnet saw reason to ask this question:

Why has the outcome of sixty years of work by many first rate scientists and at a cost of hundreds of millions of dollars had so insignificant an influence on the prevention or treatment of cancer?[6]

Despite the less than spectacular results from the NCI, President Richard Nixon—inspired by the success of the space program, which had just placed Americans on the moon—declared a national "war on cancer" in 1971. The National Cancer Act of 1971 represented, in his words, "a total commitment of Congress and the President . . . to provide the funds . . . for the conquest of cancer."[7]

Planners of the war on cancer predicted that technology would conquer cancer as it had conquered space. Molecular biology would lead the way. As the above statistics show, the battle plan of the National Cancer Program has meant little, and molecular biology and biotechnology have led us nowhere.

These facts have not been lost on members of the scientific community. In 1986, the *New England Journal of Medicine* published a study by Drs. John Bailar and Elaine Smith of the Harvard School of Public Health and the University of Iowa. Bailar and Smith are epidemiologists, scientists who study the incidence and distribution of disease. After evaluating the progress against cancer from 1950 through 1982, they concluded that

> Some 35 years of intense effort focused largely on improving treatment must be judged a qualified failure. Results have not been what they were intended and expected to be . . . *We are losing the war against cancer.*[8] (Emphasis added.)

The war is being lost because cancer research has failed to produce practical information useful in treating cancer. The epidemiologists concluded that only by switching the emphasis from treatment to prevention would the death rate from cancer be lowered. But before reaching this conclusion, Bailar and Smith asked a salient question: Why were hopes so high, and what went wrong?

The American public and many physicians are wondering much the same thing. In his 1988 presidential address to the American Radium Society, Dr. Morris Wizenberg said,

> We are not going to address the cancer issue very suc-
> cessfully as long as less than ten percent of the public
> thinks we are dealing with a curable disease; *we are also
> not going to get very far as long as large segments of the
> medical profession persist in believing that we are dealing with
> an incurable disease.*[9] (Emphasis added.)

According to Dr. Wizenberg, not only do 90 percent of Americans equate a diagnosis of cancer with a death sentence, but so too do their physicians. And Bailar and Smith have shown that the public and their doctors are not too far wrong.

In April 1991, the *New England Journal of Medicine* published another demoralizing, headline-making article about cancer treatment. Dr. Barrie Cassileth and her associates at the University of Pennsylvania studied the survival rates of patients with widespread cancer, comparing patients receiving standard chemotherapy to those receiving an unorthodox therapy regimen that is frowned upon by conventional medicine. The alternative treatment, given at a clinic in San Diego, used a strict vegetarian diet, vaccines, and regular enemas made from coffee grounds. (Every year, thousands of dying cancer patients discard their oncologists to seek the help of alternative practitioners. Let down by conventional medicine, facing certain death, hopeful patients pour billions of dollars into these clinics. Could there be a worse indictment of orthodox treatments for cancer?)

Dr. Cassileth and her colleagues found that the average survival of patients in both groups was only fifteen months. Patients in each group experienced unpleasant side effects during their treatments, caused either by the treatment or by the underlying malignancies. The researchers concluded,

> As patients and the public have become increasingly edu-

cated, dissatisfaction with conventional care for cancer has grown. The toxic effects of chemotherapy, the absence of new and markedly improved treatments despite decades of effort . . . all contribute to the dissatisfaction [of cancer patients] . . . . Our findings suggest that conventional therapy . . . should be measured against a no-treatment alternative involving only palliative care.[10]

The implications of this work were not lost on the editors of *USA Today*, the national daily newspaper, which headlined the message on its front page the same day the study was made public. "We might not be doing some of these patients any favors by treating them with chemotherapy," said Cassileth.[11]

Cassileth and her colleagues, like Bailar and Smith, are not laboratory scientists or oncologists. They are public health professionals looking at the effects of cancer and its treatments on people. Negative reports such as theirs seldom come from the cancer research or oncology community. Indeed, as you shall see, the NCI and other bastions of the cancer research establishment discourage any research that does not support the status quo, and the media follow instead of questioning the soundness of the National Cancer Program.

### GALILEO REDUX

In May 1990, I attended the annual meetings of the American Association for Cancer Research (AACR) and the American Association of Clinical Oncology (AACO) in Washington, D.C. These gatherings are the largest yearly conferences devoted to cancer research and treatment, attracting thousands of scientists and oncologists from all over the world.

One of the featured speakers at the AACR conference, which began Wednesday afternoon after the AACO conference, was Senator Ernest Hollings of South Carolina, a long-time member of the Senate Appropriations Committee, who has an important say in approving the NCI's annual budget. Hollings spoke of the unrest in Eastern Europe and the grass-roots rebellions against the communist regimes that had been in power since World War II. He commended the dissidents of these countries and praised their courage and tenacity in the face of raw power. It was a stirring speech that earned the senator a warm and enthusiastic ovation.

While listening to Hollings, I was struck by a certain irony. Though he did not know it, he could easily have been describing important aspects of the world of cancer research. In federal institutions, universities, and medical schools around the country, critical analyses of the current approach to cancer research are being squelched and ignored in favor of cell lines—an inaccurate but enormously powerful experimental model. Scientists who question this model find themselves cut off from funding, removed from academic appointments, and locked out of the ivory towers of the cancer establishment. As political dissent rages behind the former iron curtain, cancer dissent in America is effectively being silenced by some of the nation's most powerful medical and scientific organizations.

When the session adjourned, I approached Dr. Samuel Broder, the present director of the NCI, to question him about this situation. Broder oversees the institute's annual $2-billion congressional allotment. Appointed to the leadership of the NCI at the end of 1988, Broder has spent his entire professional career making his way up the bureaucratic ladder of an institute known for promoting from within.

I told Broder that the current system provided no outlet for scientists, such as myself, who disagree with the direction of cancer research. I said that I felt there should be a mechanism for the expression of legitimate dissent within the cancer research community. Broder stopped only long enough to tell me to "keep trying."

"I've tried for years," I replied.

"Well, keep trying," said the director of the National Cancer Institute; "Galileo was successful."

"Yes," I retorted, as Broder walked away. "But how many years did it take before Galileo's views were accepted?"

Galileo Galilei lived in the late sixteenth and early seventeenth centuries and is considered to be the father of modern scientific thought. He believed in the Copernican model of a solar system in which the earth circled the sun—a model that contradicted the Ptolemaic model of an earth-centered solar system. Unfortunately for Galileo, the model of Ptolemy was also the official doctrine of the Roman Catholic church. Galileo's dissent was silenced by his arrest. It took more than a century for the astronomy establishment to accept the sun-centered view of the solar system.

Although Broder's mention of Galileo was undoubtedly flippant, things have not changed much in the four centuries since Galileo's era. The National Cancer Institute has become much like the church of the early seventeenth century, afraid of losing power by having its beliefs and favorite model discredited.

The church's model was favored because it put earth and man at the center of the universe. The NCI's model is favored because it allows the technology of molecular biology to be conveniently employed. The theories and methods of molecular biology have become the cancer establishment's Bible. Dr. Broder, for all his credentials, is little more than a politician trying to maintain the beliefs, status, and funding of a huge bureaucratic organization.

But even Broder is aware that something is inherently wrong with the battle plan of the cancer war. A mere two months before my encounter with him in Washington, I listened to him speak in Tucson, Arizona, at a conference on chemotherapy. There, in the presence of his peers, the director of the NCI came close to being completely truthful about the realities of cancer research.

Broder began his talk by asking, "What are we accomplishing at the National Cancer Institute?" Then, for the better part of an hour, he admitted that the institute had produced little of practical value for oncologists. Stating that "cancer statistics are the law," Broder noted that the age-adjusted death rate for all cancers is increasing and confessed, "We are losing ground in many types of cancer." Thus, the conclusion that Bailar and Smith had reached four years earlier has been confirmed by the establishment. But Broder also told the audience of oncologists that when he is called to testify before Congress to justify the budget of the NCI, he will argue that "we are on the right track." Like the pope of 400 years ago, Broder must maintain the status quo. He has his career and a very large establishment to protect.

These were the words of the director of the National Cancer Institute, the current commander in chief of a twenty-year-old war of attrition. If we ever hope to win this war and make truly significant inroads against the modern scourge of cancer, the establishment must be willing to acknowledge its mistakes. Researchers must turn away from petri dish "cancer" to the realities of human cancer. Four hundred years ago, people were told that the sun moved around the earth. Today, the public understanding of cancer is just as inaccurate, for much the same reason: An incorrect model of nature dominates the thought within an entire scientific field.

But this model, unlike the universal model of Ptolemy, has a large impact on human health and well-being. Lax governmental regulation of the cancer industry permits what is possibly the greatest lapse of scientific judgment in history to continue. In the recent past, lax governmental regulation was responsible for the spectacular collapse of the savings and loan industry. In each instance, society pays a high price for an industry's misjudgments. But unlike the S and L mess, no amount of money can save us from the consequences of cancer.

# 2

## The Making of a Cancer Scientist

*Unfortunately, most basic scientists have little knowledge of pathobiology or clinical medicine.*[1]

Irwin M. Arias, M.D.,
Tufts University School of Medicine, 1989

My feelings about the state of cancer research are far more than the disillusionment of a frustrated research scientist. I came to this point after many years of work as a cell biologist and then as an experimental pathologist. During those years, I developed a tremendous respect for the goals of scientific inquiry. The purpose of all science should be to gain a greater understanding of the natural world and its workings.

For as long as I can remember, I have been interested in biology. Perhaps it was a curiosity about living things that prompted my parents to give me a microscope when I was about twelve years old. From my first glimpse of abundant life in a single drop of gutter water, I was hooked.

Around the same time, I read Paul De Kruif's classic book *The Microbe Hunters*, a collection of stories about early microbiologists. I devoured these stories of great medical pioneers:

Anton van Leeuwenhoek, the Dutchman who invented the microscope; Robert Koch, the German physician who discovered the tuberculosis bacterium; and Louis Pasteur, the French chemist who discovered the role played by microorganisms in fermentation and disease and developed a curative treatment for rabies. These men became my boyhood heroes. They had used instruments more primitive than mine to uncover central truths about human disease, often in the face of tremendous opposition. From the moment I opened De Kruif's book, my fate was sealed. I would be a scientist.

## UCLA

When I entered UCLA in 1956, I wanted to become a physician. After a single year of premedical studies I realized that medical practice would not be the world of experimentation and discovery that I had envisioned. Physicians were more technicians than scientists, using their skills to apply knowledge gained elsewhere—in the laboratories of researchers. I wanted to be one of the discoverers of that knowledge, so I abandoned my premedical studies to seek a career in research.

At the start of my sophomore year, I switched majors from premedical to biophysics. Although a biophysics degree would give me an excellent background in science, it was also one of the most difficult majors to complete. I would have to take chemistry courses for chemistry majors, physics courses for physics majors, biology courses for biology majors, and math courses for math majors. Every class would be challenging, but I wanted a broad understanding of the sciences. I completed my undergraduate studies in biophysics in four and a half years, followed a year and a half later by a master's degree in genetics.

I had decided on the direction of my doctoral studies even

before I had earned my master's degree. I wanted to focus on the most basic unit of life, the cell. Cells are the indivisible structural and functional building blocks of all living tissue. Life began about 3.5 billion years ago as simple, single-celled organisms called prokaryotes. Bacteria are examples of unicellular prokaryotic organisms living today. Larger, more complex eukaryotic cells containing nuclei and other internal structures called organelles appeared some 1.4 billion years ago. Common unicellular eukaryotes are the amoeba and the paramecium. Multicellular organisms made up of large numbers of eukaryotic cells arose only about 600 million years ago. We are a modern example.

The condition of cells, the efficiency with which they carry out their tasks, determines the health of the entire organism. All disease is the result of damage to, or changes within, individual cells.

Despite their fundamental importance to the quality and continuation of human life, there is still much we do not know about cells and their functioning. By today's standards, scientific knowledge about cells was almost rudimentary when I began my graduate studies in 1961. The molecular structure of DNA, the material of genes, had been known for only eight years. The mechanisms by which genes are turned on and off in cells were just beginning to be understood. I wanted to be part of a new frontier, so I pursued a doctorate in cell biology.

Graduate school was more exciting than undergraduate studies, largely because I could concentrate on research. I developed research skills by emulating more experienced people in the laboratory, such as the more advanced graduate students to whom I always went for advice. I was part of a supportive community of researchers, all working to discover new information about cellular processes. It was in this context that I met the man who would leave an indelible mark on my professional career, Professor Fritiof Sjostrand.

## SJOSTRAND

Dr. Sjostrand, who was a medical school graduate, had abandoned the practice of medicine for research. He had been recruited by UCLA from the Karolinska Institute in Stockholm, Sweden, in 1959. He was a pioneer in the use of the powerful electron microscope, a large instrument that can magnify objects to more than 100,000 times their normal size. Unlike conventional microscopes, electron microscopes use electron beams rather than light, making them more than 100 times more powerful than the most advanced light microscope.

Sjostrand was one of the first researchers to use the electron microscope to study the makeup of individual cells. During the 1950s he had harnessed the new technology to examine the structure of cell membranes and the organelles within cells, such as mitochondria and the endoplasmic reticulum, improving science's understanding of cell structure and function. During the 1960s it was widely believed that Sjostrand would receive the Nobel Prize. Although this never happened, he did receive numerous international awards for his trailblazing research on the ultrastructure of cells, and his presence at the university was a great coup.

In 1961, Sjostrand selected me to be one of his graduate students because of my broad scientific background, giving me the opportunity to learn electron microscopy from one of the field's foremost experts. In Sjostrand's lab, I found my niche and developed the skills and knowledge of cells that would later open my eyes to the scandal of cancer research.

When I joined Sjostrand's group at UCLA, he had already assembled a team of more than thirty people. Some were visiting scientists from Sweden, others were technicians, and still others were graduate students. Sjostrand had an active and restless mind and was involved in diverse research projects. Graduate students were given their pick of these projects to work on for their doctoral dissertations. I chose to investigate the absorption of lipid (or fats) by the cells of the mammalian intestine, using laboratory rats as my model.

The experience of working in Sjostrand's lab had a tremendous effect on the development of my scientific values concerning the way research should be conducted. I enjoyed the camaraderie and support of a communicative and often argumentative scientific community where ideas were openly discussed and debated. The contrast between Sjostrand's lab and the insular, protectionist world of cancer research that I later entered would be a source of continuing dismay to me.

Although Sjostrand was an incredibly busy man, juggling his research and teaching commitments with editorial duties on a scientific journal, he would join the laboratory staff every afternoon for an extended coffee break/discussion session. We would talk about the progress (or lack thereof) of our own research and of the recently published findings of other labs. We would all offer and accept suggestions and generally benefit from the feedback. For a young researcher in training, it was an incredibly exciting and educational time.

In addition to the hours spent in the laboratory, I frequently got to see Sjostrand in action at more public occasions. Although he was only in his early fifties at the time, he could have the passion and conviction of a biblical patriarch when defending his ideas. He would get up at lectures to criticize vigorously the work of others who did not agree with his views on the structure of cell membranes. At the time, I was a little embarrassed by his outspokenness. Now, some thirty years later, I find myself in a similar situation in the world of cancer research. But, unlike Sjostrand, I have few (if any) scientific venues where I can express my views.

## THE HOSPITAL OF THE GOOD SAMARITAN

At the end of graduate school in 1966, I received a fellowship from the National Institutes of Health that allowed me to spend two years in a biochemical research laboratory at the University of Lund, in Sweden. While there, I continued the work on lipid absorption. When the grant ended, I returned to the United States. Years of research had made me the epitome of a basic scientist, one with a passionate interest in learning how cells work but very little awareness of the practical, health-related aspects of biological research. I was also expert at a rather new technology, electron microscopy.

By the 1960s, pathologists realized that the electron micro-scope had a role in surgical pathology. For this reason, the Hospital of the Good Samaritan in Los Angeles decided to establish a laboratory of electron microscopy and was looking for a director. I learned through a colleague of the position, and although I had no knowledge of or interest in pathology, I applied for the job and was hired.

The chief of pathology at Good Samaritan was Dr. William Kern, an outstanding surgical pathologist who was also a clini-cal professor of pathology at the University of Southern Cali-fornia School of Medicine. When I accepted employment, Kern knew what I had yet to learn—that in about 99 percent of cases, pathologists could arrive at definitive diagnoses by using con-ventional light microscopes. Kern's goal was to establish a state-of-the-art facility where the electron microscope's in-creased magnification power could provide accurate diagnoses in the few difficult cases for which the light microscope was inadequate. He knew that my clinical responsibilities would be moderate and encouraged my interest in research. It was an ideal situation for us both, and I was glad to move back to southern California.

The pathologist's knowledge of the structure and behavior of cells has real impact on the lives of patients. It is the patholo-gist, not the surgeon or oncologist, who determines if a lesion is a tumor, and whether it is malignant or benign, and who casts the final, decisive diagnostic vote. All medical specialties rely on pathology for an accurate diagnosis of disease. Pathologists do their job remarkably well because they evaluate under a microscope the characteristics of the basic units of life itself. Different diseases cause different alterations of the body's cells.

Shortly after arriving at Good Samaritan, I began attending noon conferences, wanting to get a better understanding of my new job. At certain conferences, a physician would present a particularly interesting and difficult case that was not his or her own. Using data from the patient's records, the physician would explain the results of the physical examination, x-rays, and other diagnostic tests, finally deciding on the most likely diagnosis. The audience of physicians would then discuss the case, sometimes offering other opinions. Finally, a pathologist would project some slides of tissue sections from the patient, as seen through a microscope, and then end the suspense by divulging the correct diagnosis. At conference after conference, the definitive diagnosis was based on the pathologist's microscopic examination of the patient's cells.

Although the electron microscope provided much higher magnification than the standard microscope, pathologists really did not need the extra magnification to do their jobs most of the time. In the rare hard-to-diagnose cases, of course, the electron microscope was an invaluable tool. But between such cases, I had ample time to conduct my own research. I had only to choose a topic.

## CANCER

Before joining the staff of Good Samaritan in 1969, I had used laboratory rats as my experimental model for mammals in general. As an acute-care medical center, Good Samaritan did not have animal facilities. If I was going to continue to do research, I would have to use human material, tissue specimens from patients undergoing surgery at the hospital.

In 1969, President Nixon had not yet declared war on cancer

and I myself knew nothing about the disease. The surgeons of Good Samaritan did not share my inexperience, since a number of cancer surgeries were performed there every day. The importance of the cancer problem plus the availability of human material made cancer a logical choice for research. I began a literature search, both to assess the state of knowledge in the field and to find an interesting topic for experimentation. I simply dove in, blissfully unaware of my own naiveté.

Cancer is an ancient disease. The fossil record shows that cancer has occurred in animals for tens of thousands of years. The physicians of ancient Egypt, Greece, and Rome documented cases of cancer in their patients. Then, as now, it confounded and frustrated physicians. The Greek physician Hippocrates (460–370 B.C.) observed cancer of the skin, breast, stomach, cervix, and rectum. In a breast cancer patient, Hippocrates described distended veins radiating from the tumor that looked like a crab's claw. He then named the disease *cancer*, which is "crab" in Latin. Hippocrates believed that tumors deep in the body should not be treated because if they were, the patients would die quickly. Ironically, today, more than two millennia later, some experts in the field, like Dr. Barrie Cassileth, are presenting similar arguments—that is, withholding treatment is as effective as the best available drugs.

Knowledge about cancer continued to accumulate in Rome and Alexandria, where operations to remove tumors were performed. The physician Galen (130–200 A.D.) believed that cancer was caused by an excess of black bile. This incorrect theory influenced medical thought for more than 1,000 years.

By the eighteenth century, the influence of Galen had waned and physicians began to note that cancer began as a local disease and later spread through the lymphatic vessels to the lymph nodes and then into the general circulation. The spread of cancer to other organs of the body was explained by this new concept.

An English surgeon named Percival Pott (1714–1788), working in London, is credited with discovering the first link between an environmental carcinogen and cancer. He observed that young chimney sweeps had a high incidence of skin cancer of the scrotum. Pott deduced that the repeated application of soot to the scrotal skin of these adolescent boys as they were thrust up narrow chimneys caused the development of cancer, particularly since the boys seldom bathed and the soot remained on their bodies for long periods of time.

The work experience of these boys turned out to be an unfortunate, unplanned experiment on cancer and the environment. The experiment showed that chemicals can cause cancer. By 1915, scientists were producing cancer in experimental animals by repeated applications of coal tar to the skin. By the middle of the twentieth century, the public was being warned about the carcinogens in tobacco. Also by the end of the nineteenth century, scientists were beginning to understand that hormones, such as estrogen and testosterone, are closely linked to some forms of cancer.

The modern era of pathology began with the German pathologist Rudolph Ludwig Virchow (1821–1905). In an 1858 publication, he established the cell as the all-important unit on which and in which all disease processes are active. During the 1870s and 1880s, with the refinement of microscope lenses, rotary microtomes that cut very thin slices of tissue, and aniline dyes that stain nucleus and cytoplasm different colors, Virchow's vision of a cellular pathology became a reality. Static images of cells under the microscope continue to provide the most reliable evidence that distinguishes one disease from another or from normal tissue.

Cancer begins when one of the 100 trillion cells that make up the body becomes defective and stops obeying the rules that govern normal cell behavior. This cancerous cell does not develop, age, and multiply normally. The mechanisms that tell a normal cell to stop dividing are not functioning. Instead, cancer cells grow and multiply, eventually forming a mass of cells known as a tumor. As the tumor grows, it presses on the cells surrounding it, causing damage. Cancer cells can also invade neighboring tissues, sprouting extensions that reach like roots into the layers of tissue surrounding the original mass.

Some tumors form discrete, well-defined masses that do not continue to grow. These benign tumors (which include freckles and moles) are markedly different from the malignant forms of cancer that strike fear into the hearts of patients. Malignant cancers continue to grow and have the ability to spread to other parts of the body, both through direct invasion and the deadly process known as metastasis.

Metastasis occurs when some of the cancer cells break away from the original tumor and are carried by the lymphatic and circulatory systems to other organs of the body. There they begin to grow again, forming secondary tumors. The body is defenseless against these abnormal masses of cells and eventually succumbs when one or more vital organs is destroyed.

Though most cells can become cancerous, some types are more prone to cancer than others. The most common forms of cancer are carcinomas, which arise in the cells that form the linings or coating of various body organs. Most cancers of the lung, breast, prostate, and colon, for example, are carcinomas. Cancers that arise from other tissues, such as connective tissue, muscle, or bone, are called sarcomas. Cancers that form from blood cells are known as leukemias or lymphomas.

Cancers are classified on the basis of their differentiation—the degree to which the malignant cells resemble the cells of the organ in which they originated. The cells of a well-differentiated pancreatic tumor, for example, when examined under a microscope, closely resemble normal pancreatic cells. Poorly differentiated tumors, in contrast, show less resemblance to the cells of their "parent" organ and also usually grow and spread at a faster rate than well-differentiated tumors. For these reasons, poorly differentiated tumors are often referred to as "aggressive" cancers.

Cancer can be caused by many factors in our external and internal environments. But it is best understood as a single disease that appears in many varieties, depending on which of the 200 body cell types is affected. Cancer is a single disease because every malignant cell breaks the same rules of normal cellular behavior and, as we shall see, shares certain other traits with every other malignant cell. However, just as every human being is unique, so every individual cancer has its own unique, if subtle, features.

## EXPERIMENTATION

Scientific experiments are designed to answer a specific question concerning a disease process or natural phenomenon. After reviewing the cancer literature, I saw clearly that one of the most pressing questions for cancer scientists is this: Why are malignant cells able to move around in the body when normal cells cannot? Cancer metastasis was (and is) a core challenge for cancer research.

Then, as now, it was believed that the surface membranes of malignant cells were altered in some way that allowed them to detach themselves from the tumor and travel to other organs. This theory was interesting to me after my years of work on the cell membranes of absorptive cells in the intestine. I was drawn to the metastasis problem and decided to get my feet wet in my new field by investigating the surface properties of malignant cells to determine what made them different from nonmalignant cells.

If my first experiments produced data that supported the cell membrane theory of metastasis, I would conduct additional experiments to test the theory more vigorously. However, if my initial experiments produced data that did not support the theory, I would either design different experiments or decide very quickly that the theory was wrong. I could then propose a new theory to explain metastasis. My goal, like that of any scientist interested in the problem, was to collect and analyze information until the mechanism of metastasis could be explained.

This plan followed the most basic principles of the scientific method, a twofold procedure of theory and experimentation that was pioneered by Galileo and his contemporary Sir Francis Bacon. Both insisted that scientific research must rely on observation. The goal of all experimentation is a greater understanding of the workings of nature, based on careful observation of the natural world. It is not enough to "think" or "believe" that a theory is correct—it must be proved through experimentation.

With accurate observations in hand, a scientist meditates on a problem by using logic, analogy, and imagination. Scientific creation occurs when facts are connected in a new way and a novel idea is formed for further exploration. If this new idea is correct, it will increase knowledge of the world and accurately predict its behavior. Einstein's theory of special relativity, for example, predicted the atomic bomb and nuclear energy. An inaccurate theory will not predict natural processes.

In medicine, the success of the scientific method is largely dependent on the accuracy of the models used for experimentation. Since only very limited experimentation can be performed on human beings, biomedical scientists must use *experimental human models* in their work. These experimental stand-ins enable scientists to study disease processes under controlled laboratory conditions. In the laboratory, the experimenter can separate from the complexity of a disease a few limited aspects that he or she chooses to investigate. The accuracy of the model can be judged by how well the experimental findings match what is actually observed in humans with the disease.

Over the centuries, the scientific method has proved to be perhaps the greatest of all tools. Experimental science led to the technologies of modern civilization. Application of the scientific method to the problem of cancer metastasis should therefore lead to both understanding and controlling it—if correct experimental models are used.

# 3

---

# The Road to Dissent

*He is less remote from the truth who believes nothing, than he who believes what is wrong.*[1]

Thomas Jefferson,
*Notes on Virginia*, 1781

*I have realized from talking to pathologists that cancer is more than just a bunch of cells in a petri dish.*[2]

Pierre Chambon, molecular biologist, 1991

My choice of an experimental model for cancer research was obvious from the start. The hospital conducted cancer surgery every day, and I could obtain tissue samples from hundreds of tumors every few months. Some would be poorly differentiated, with high metastatic potential, whereas others would be well differentiated, with low potential. My first research task was to demonstrate that the cell surfaces of poorly differentiated tumors differed from the cell surfaces of well-differentiated tumors.

To test my hypothesis, I decided to use special stains that would bind to a specific sugar, called sialic acid, which is found on cell surfaces as a component of large glycoprotein molecules. Glycoproteins are proteins to which various amounts and kinds of sugar molecules are attached. I applied the stains to thin tissue sections and then examined the sections in the electron microscope.

## EARLY FINDINGS

My first experiments concerned breast cancer and the normal breast tissue that is always found in mastectomy specimens. (Those observations were published in 1973.)[3] When I examined the stained sections under the microscope, I observed that the surfaces of malignant cells were less stained than the surfaces of normal cells. Malignant breast cells seemed to have glycoproteins that contained less sialic acid than normal breast cells. I decided to expand the project and see if other types of cancer displayed similar chemical alterations in their cell membranes.

The next study was carried out in collaboration with Dr. William Kern, the hospital's chief pathologist. Kern had a long-term interest in bladder cancer, so we decided to study it and the normal bladder, portions of which are often resected with bladder tumors. Once again, specimens were obtained, tumors were classified according to their degree of differentiation, and each specimen was prepared and stained. And once again, the cells from poorly differentiated tumors exhibited less stained material on their outer surfaces than the normal bladder cells or the cells from well-differentiated tumors. In publishing the results of this work, Kern and I proposed that the loss of cell membrane sugars contributed to a reduced "stickiness" of tumor cells, allowing them to metastasize.[4] In 1989, another laboratory, using different methods, confirmed our findings.[5]

In considering the surface properties of malignant cells, I was struck by their similarity to normal cells that are at the end of their life cycle. Old cells also fail to "stick" to their neighbors. In the bladder, for example, old cells fall off the bladder wall into urine, where they die. In the intestinal tract, old cells slough off into intestinal contents. I was sufficiently curious about this similarity to return to the electron microscope to search for old bladder cells that were about to be sloughed off into urine. In these unpublished observations, I found that the surface glycoproteins of normal bladder cells seemed to contain fewer sugars as they aged. Others had already demonstrated that the surfaces of red blood cells lost sugars as they aged. It appeared that both the surfaces of aggressive malignant cells and those of old normal cells were "nonsticky," and for the same reason.

It seems logical that some parts of the biochemical machinery within cells would slow or even stop as part of the normal aging process. The chemical reactions that add sugars to proteins could well be one of the processes affected by cellular aging. In normal old cells, this change would be harmless because old cells have also lost their ability to divide. But if events related to aging occurred in a malignant cell that was multiplying uncontrollably, it could become the most frightening of cancer cells—a malignancy with the potential for metastasis.

By experimentation with human tumors, I had come upon a reasonable explanation of the metastatic process, but few of my colleagues in the field were interested in my observations. Their studies pointed in other directions. Some even believed that malignant cell surfaces gained sugars, a finding opposite to mine. I began to wonder why.

In 1974, the same year that Dr. Kern and I published the paper on bladder cancer, I conducted another investigation of breast cancer, this time in collaboration with Russell Sherwin, M.D., professor of pathology at the University of Southern California School of Medicine. What we found would irrevocably change my view of cancer research.[6]

## THE TURNING POINT

Normal breast cells are secretory. In nursing mothers, the cells of the breast make and release a specialized secretion (milk) into tiny tubes called milk ducts. Eventually, larger ducts carry the milk to the nipple. What most people do not realize is that breast cells exhibit low levels of secretory activity all the time. This is why many nonnursing women, with a breast pump, can express drops of milklike fluid from their nipples. The cells of the breast are maintained in a partially developed or differentiated state, and as a result they exhibit a continuous low level of secretory activity. The experiments that Sherwin and I conducted showed that when breast cells become malignant, they continue to produce secretions.

I was completely taken aback by this finding. Virtually everything I had read in the research literature indicated that cancer cells are profoundly different in behavior from the normal cells of the organ in which they arise. Yet my experiments on breast tumors from actual cancer patients showed that malignant breast cells looked and behaved much like normal breast cells. The differences between normal and malignant were subtle, not profound.

Although our results differed widely from the accepted dogma concerning cancer cells, Sherwin's and my findings fit well with what pathologists have long known about cancer. Every day, in pathology labs around the nation and around the world, pathologists described the fixed, differentiated features of human tumors.

In the early days of my research on cancer, it was hard for me to believe that the experts who were publishing in cancer research journals were wrong. Perhaps my research had been done sloppily. With each experiment, I tried to be more meticulous in my methods and analyses. I read more books and journal articles about cancer. I attended the yearly national cancer research conferences. I spoke with scientists from all parts of the country and around the globe. And I gradually realized that the work of most cancer scientists and the main direction of cancer research, in general, was pointing to an understanding of malignancy very different from mine. Very little information in experimental cancer journals was compatible with what I was learning from careful experiments with human tumors and from working side by side with hospital pathologists.

Experiments that I conducted on human lung tumors in 1980 provided yet another example of the fundamental differences of opinion that separated me from most other cancer scientists.[7] This final research project at Good Samaritan concerned the cellular origin of lung cancer.

There are two bronchial tubes, divisions of the trachea, that carry air into the right and left lungs. The inner walls of these tubes are lined by several layers of cells and are part of what is called *respiratory epithelium*. There are six different types of cells in respiratory epithelium, each having a distinct microscopic structure and function. Most lung tumors arise from these epithelial cells.

Lung tumors are also of several different types, and in these final experiments I investigated which of the six cell types gave rise to a common type of lung tumor known as an adenocarcinoma. I took small pieces of resected, living lung tissue—containing portions of the normal bronchial wall and/or adenocarcinomas—and placed them in a liquid medium containing sugar molecules to which a radioactive isotope of hydrogen had been attached. Cells utilize the radioactive sugar and the common, nonradioactive form in the same manner. The radioactive sugar molecules entered into the cells in the lung tissue specimens and became incorporated into parts of the growing normal and malignant cells, just as the normal sugar would. The newly manufactured components were identified by their radioactivity.

A comparison of the normal cells of the bronchial wall and the adenocarcinomas showed that some of the parts made by the tumor cells were found in only one of the six normal cell types—ciliated cells. Cilia are the microscopic hairlike processes that are found on these cells. In the bronchi, cilia project from cell surfaces into moist airways, where, under a covering of mucus, they wave back and forth to move particles out of the airway and into the throat, where they can be swallowed.

The tumor cells, although they exhibited many of the characteristics of ciliated cells, did not actually have cilia. The malignant cells appeared to be developmentally "blocked" and could make only some of the parts for cilia. As a result, cilia were lacking. My experiments revealed that the cellular origin of lung adenocarcinomas were cells of the bronchial wall that were in the process of developing into mature ciliated cells. I concluded that carcinogenesis alters the normal maturation of cells in the human lung.[8]

Once again, my research did not match the bulk of information in the cancer research literature. Most of the experts in the field are unaware of the close ties between cellular developmental processes in organs and cancer's beginnings.


## THE LIGHT DAWNS

After ten years of doing experiments, reading books and papers, going to conferences, and listening to others, I finally pieced the puzzle together and realized why my understanding of cancer was very different from that of the recognized authorities. Cancer research is an experimental science, and the most important part of any experiment, the foundation on which everything rests, is the organism that is chosen for study. I had spent twelve years at Good Samaritan Hospital working with pieces of living human tumors obtained soon after surgery. The malignant cells in the surgical specimens were little changed from the time they were living in patients. I understood from the beginning that careful experiments with living tumors would provide the most reliable information about how cancers behave in the body. I assumed that all cancer researchers tried to work with good representations of human cancer *in vivo*.

As I noted in Chapter Two, there are poor experimental models as well as good ones. In cancer research, a good model is one that is closely related to human cancer. Information drawn from such a model is likely to increase our understanding of the disease. A model that is not closely related to human cancer will not increase our understanding. Like the geocentric model of the solar system, a poor experimental cancer model will foster misleading and incorrect beliefs about nature.

After a decade in the field, I realized that most cancer researchers do not employ a good representation of human cancer. Instead, they select a model based on faith alone, because it is convenient to use and allows them to publish a maximum of research papers. Since success in the field comes primarily through these papers, cancer researchers know they must "publish or perish" if they are to survive in their careers. To meet their career goals, most rely on an easily studied and manipulated model of cancer, known as *cell lines*.

I had learned a great deal about cell lines by reading extensively in the field and by discussing their characteristics with many scientists who work with them. I saw clearly that a lack of shared knowledge between the fields of pathology and cancer research had impeded real progress against the disease and that my best hope of changing the situation was to become more involved with the world of academic cancer research. As a researcher working in a clinical setting, I was cut off from my peers in research. I was able to discuss research issues only when I attended conferences. I longed to have regular discussions about cancer with other scientists and to convince them of the folly of using cell lines as experimental models. With this goal in mind, I left the Hospital of the Good Samaritan at the end of 1980 and accepted an appointment in the medical school of the University of North Carolina (UNC) at Chapel Hill.

## THE MOVE TO NORTH CAROLINA

UNC's medical school consistently ranked high in the amount of grant money awarded by the NCI. It has a multimillion dollar cancer research center, built with federal funds and private contributions. Many of the university's cancer scientists work at this center.

Soon after my arrival in Chapel Hill, I went to the chief of surgical pathology at the university hospital to set up a procedure for obtaining samples of human tumors. He looked at me strangely and asked what I wanted them for. I told the pathologist that I had just moved and wanted to continue my research. He gave me access to excess surgical material that was not needed for diagnostic purposes but continued to question me. No other scientist at the university or medical school had ever made such a request. Hundreds of cancer researchers worked there, but not one of them studied human tumors. At that moment, the essential nature of the differences that separated me from the other scientists at UNC hit home.

It had been my good fortune to begin a cancer research career in a hospital, where the only available research material was human cancer. Pathologists diagnosed it, surgeons cut it out, oncologists treated it, and it was discussed in detail at weekly meetings. I knew how human cancer cells behaved because I had spent more than a decade working with actual human tumors and with the clinicians who deal with cancer patients.

In an acute-care hospital such as Good Samaritan, the goals of all the professionals are practical. From the nurses at their stations to the pathologists in their offices, the common goals are the diagnosis and treatment of disease. I found out, too late, that researchers at a medical school have very different priorities and operate in a very different environment. Their laboratories, and the work that goes on inside them, are far removed from human cancer patients and their malignancies.

When I arrived at the Hospital of the Good Samaritan I was

a cell biologist. When I left twelve years later I had become an experimental cancer pathologist—a scientist who investigates the structural and functional changes in human cancer cells. Most cancer scientists at academic institutions consider themselves molecular biologists. Molecular biologists focus their attention on the molecules within cells.

Molecular biology's "golden age" began in 1953, with the discovery of the molecular structure of deoxyribonucleic acid (DNA). From that time on, molecular biology should really have been called molecular genetics, because most molecular biologists are interested in only one type of molecule within cells—nucleic acids, the material of genes. Molecular biologists have developed powerful methods for studying DNA and ribonucleic acid (RNA), and they are deeply wedded to this technology.

The methods of molecular biology are popular because they produce a great deal of information in a short amount of time. By breaking apart cells and their genetic material, scientists can produce volumes of information that can be discussed endlessly in papers. To perform these feats they require cells that are plentiful, easy to manipulate, and easy to study in their research laboratories. They therefore choose experimental models that fit these criteria—like cell lines.

Cell lines are cells that have been grown in culture, on the bottoms of round, flat plastic dishes. They are the distant descendants of actual human or animal cells that were removed from bodies and transplanted to culture many years before. Cells that are thus established in culture divide every couple of days, doubling the amount of their DNA in the process.

Most cancer scientists and molecular biologists start to work with cell lines as graduate students, following the lead of their mentors. Few ever question the suitability of their research model. Their standard refrain is that "cell lines allow me to do things I cannot do in any other way." This argument might hold water if the information produced by cell line research were relevant to humans.

When cancer is reduced to clumps of cells in a petri dish—
and then further reduced from these cells to single molecules
within them (DNA)—it is inevitable that an accurate picture of
the disease will be irretrievably lost. Most cancer scientists not
only are missing the forest for the trees, but they are in the
wrong forest.

## DISILLUSIONMENT

When I moved to the University of North Carolina, I had
assumed that I would be able to collaborate and discuss my
observations about human cancer with other researchers in an
open-minded scientific environment. I believed we were all
working on the same problem, toward the same goal—an understanding of cancer that could lead to its cure or prevention.
I was wrong.

The faculty at UNC discouraged conversations that were
critical of cancer dogma. My attempts to inject controversy into
seminars, lectures, and one-on-one discussions were met with
silence or derision. At one lecture I was told to sit down by
several faculty members in the row behind me after I asked a
question that was considered controversial. I was soon labeled
a troublemaker and viewed as a second-class scientist because
my background in cancer did not include a prominent mentor,
nor had I worked at a famous research institution.

My experiences at UNC were a far cry from my experiences
in Dr. Sjostrand's lab. Instead of an open, inquisitive scientific
community, I found myself in the midst of a competitive and
suspicious group of scientists who were scrambling to reach the
top of the academic ladder. For many of them, cancer research
was a means to an end, and the acquisition of accurate or useful
information was not even a secondary goal.

One researcher went so far as to tell me that the fact that he
had a family to support was ample justification for his use of
cell lines as an experimental model. A young pathologist who

was writing a grant application to the NCI for research funds freely admitted that he did not think the proposed project applied to human cancer, but he wanted to "get on the bandwagon." An up-and-coming researcher flatly informed me that all a scientist needed to become successful in cancer research was an attention-grabbing gimmick. A few years later, this same researcher received a prestigious award from the American Association for Cancer Research for his gimmick—which applied cell lines to the study of metastasis.

During my time in Chapel Hill, I was attracted to the principles of the famous French philosopher, scientist, and mathematician of the seventeenth century, René Descartes. Descartes was convinced that much of accepted knowledge is inherently unreliable and that we should view the truthfulness of most of what we are told with doubt rather than faith. The truth is something that must be proved before it is believed. This is also a very American approach, as exemplified by Thomas Jefferson, a fellow skeptic and doubter who wrote that "he is closer to the truth who believes nothing, than he who believes what is wrong."

Jefferson's devotion to accuracy and truth is evident in his 1781 book, *Notes on Virginia,* in which he also showed himself to be an able naturalist. Jefferson took European naturalists to task for assigning characteristics to the animals of America without actually seeing them. He noted that close examination and questioning of the methods of European experts would "probably lighten their authority [and] render it insufficient for the foundation of hypothesis."[9] Jefferson's skepticism seemed singularly appropriate to the world of cancer research in which I found myself at the University of North Carolina—where researchers made hypotheses about cancer without ever having studied a real human tumor.

I was even more isolated in the research environment of UNC than I had been at the Hospital of the Good Samaritan. My colleagues were not interested in hearing my views, and I ended up working exclusively with one other scientist, Dr. Jess

Edwards, whom I came to respect greatly for his experimental skills, uncompromising pursuit of the truth, and breadth of scientific knowledge. After hours of working in my lab, I would walk down the hallway past neighboring laboratories and wonder at the extreme closed-mindedness that I had found at such an eminent institution. I could not believe that I was alone in my view of cell lines or that every researcher was as protectionist and narrow-minded as those I had met in Chapel Hill.

In the first half of 1983, two letters appeared in *Science,* the Journal of the American Association for the Advancement of Science, that raised my hopes about the state of cancer research. Two scientists expressed the view that cancer research was making little progress toward an understanding of malignancy. It was encouraging to see letters that expressed views similar to my own in the most widely read magazine for scientists. However, I felt that neither writer had gone far enough in his critique. I wrote my own letter to *Science*—a direct attack on cell lines, cancer dogma, and the entire cancer establishment. After two years as a pariah at the University of North Carolina at Chapel Hill, I was pulling no punches.

> Letters from Harry Rubin and Philippe Shubik have expressed the belief that recent advances in molecular biology have not led to a deepened understanding of the nature of malignancy. I would go one step further and say that some cancer research may actually have set back our understanding because the data produced are not relevant to the human disease. The pressures of decreasing research funds, intense competition and furthering careers often make publishing more important than meaningfulness of data. Thus, it is not surprising that long-term "normal" and "transformed" cell lines are favored systems for study, as data can be quickly obtained. The cells grow rapidly, producing large numbers in a short period....[10]

## THE FALL

*Science* published my letter in the summer of 1983. Not long afterward, I attended a major cancer conference that was held at the Cold Spring Harbor Laboratory on Long Island, New York. Although the four-day conference was titled "The Cancer Cell," few of the papers had anything to do with human cancer. I had several conversations with featured speakers, during which I noted that it was misleading to extrapolate experimental results from cell lines to human cancer. For the most part, I was either rebuffed or ignored.

During the course of a conversation with Dr. Helmut Land, a postdoctoral fellow from the Massachusetts Institute of Technology, I found out why so few of the several hundred conference participants were interested in speaking with me about cancer. According to Dr. Land, "It is because you are too controversial." I was surprised. I did not know any of the scientists attending the conference, and my own experimental work in pathology was little known by molecular biologists. Why should I be controversial?

Then I remembered the letter I had written to *Science*. By attacking cell lines, the favorite cancer model, I had also attacked everything about cancer research and cancer researchers. If their favorite model was wrong, then all the information and ideas that have flowed from it were also wrong. The work of tens of thousands of scientists, over a period of more than thirty years, would be invalidated. The cell line model of cancer is like the foundation card in a house of cards—pull it out and the entire "house" of cancer research will come tumbling down.

At the Cold Spring Harbor conference I found out what a career as a dissident cancer scientist would be like—virtually impossible, because I would be shunned by most of my peers.

Several months later, I received a letter from the chairman of my department at UNC. He informed me that my contract would not be renewed at the end of the year. Although I was not pleased to receive the letter, I cannot say that I was surprised. My chairman was well known for his work on cell lines and my criticism of them at the university and in the *Science* letter had probably embarrassed him.

Millions of discoveries have resulted from cancer research during the last three decades, but most of them share one common and fatal flaw. They were all based on cell lines, a terrible cancer model that shares few characteristics with the human disease. The continued reliance on this useless cancer model in thousands of laboratories around the world is a scandal whose scope would be hard to match in the history of medicine or science.

# 4

# The Immortal Cell

*Remarkable spontaneous changes during or after establishment of mammalian cells in continuous culture have been reported by several laboratories . . . . If cultural study of the cellular basis of neoplasia is to be effective, it appears important to distinguish artifactual from intrinsic sources of such cellular variation.*[1]

J.J. Clausen, Ph.D., and J.T. Syverton, Ph.D.,
Department of Microbiology,
University of Minnesota, 1961

Since the early part of this century, investigators have attempted to grow cells from human tumors in their laboratories. Their goal was to develop a convenient experimental model in which to study the disease. By studying living cancer cells *in vitro*, or outside the body, they hoped to obtain information that would be useful in the treatment of cancer *in vivo*, or inside the body. A variety of methods were tried, but none was successful—until 1951.[2]

That year, a woman entered Johns Hopkins Hospital in Baltimore, Maryland, for treatment of cancer of the cervix. Although she later succumbed to her disease, the patient, Helen Lane, has since achieved an odd measure of fame and immortality. Cells from her cancerous cervix were removed, trans-

37

ferred to a petri dish, and successfully grown in culture. Thus the
first cell line was created. These cells, dubbed HeLa after their
ill-fated patient of origin, are still being used by researchers
throughout the world. Direct descendants of Helen Lane's tumor
cells remain alive today, more than forty years after her death.


## THE MAKING OF A CELL LINE

Cell lines are immortal cultures of cells derived from animal or
human tumors and normal tissues. The cells grow forever in
thousands of laboratories around the world as microscopically
thin films on the bottoms of the small, round, covered glass or
plastic dishes named after Dr. Julius Petri, an early German
bacteriologist. A fluid layer above the film of cells, called a
culture medium, provides nutrients for the rapidly multiplying
cells. They never die, and no one knows why. When a dish is
filled, the cells are scraped gently off the bottom, divided up,
and transferred to empty dishes, where they start growing
again.

Tumor cell lines such as HeLa are created by removing a
small portion of a malignant tumor and quickly transferring it
to a laboratory before many of the tumor cells have died. At the
laboratory, the tumor fragment is cut into tiny pieces, which are
placed in a petri dish containing culture medium. An enzyme
is then added to the medium that, over several hours, digests
the connective tissue that surrounds and entangles tumor cells.
Millions of individual malignant cells are released into the
culture medium as the connective tissue is eaten away.

Once the connective tissue is gone, the cancerous cells are
collected and distributed into petri dishes containing fresh
medium. The dishes are then covered and placed in an incubator. Events occurring in the culture dishes are followed with the
use of a microscope.

Many tumor cells die soon after settling to the bottom of a

dish because they are not able to adapt to living in culture. The cells that survive flatten out and attach themselves to the bottoms of petri dishes. This is called a *primary tumor cell culture.* Over a period of months, these cells will divide a few times before they, too, die. Most tumor cells have a very limited lifetime in culture.

Occasionally, however, something else happens. A burst of cell division is detected on the bottom of a dish. The cells divide frequently, every two days, and soon fill the entire bottom of the dish, annihilating all the other cells. In some unknown way, one of the cells that should have had a limited life expectancy survives and emerges as a cell with an unlimited life expectancy in culture. Unlike its ancestors, it is exquisitely adapted to life in culture. It is immortal. It is permanent. It is established. A tumor cell line has been born.

## THE LURE OF CELL LINES

The many hundreds of cell lines that have been created can be frozen and shipped everywhere. When warmed, they come back to life. They have spawned a billion-dollar industry, made up of hundreds of companies that supply the equipment, instruments, plastic and glassware, chemicals, and media necessary to grow and study them.

Cell lines are a very convenient research tool because the cells grow rapidly. A laboratory incubator holds many stacks of petri dishes, each dish containing millions of cells, growing and dividing every few days. In this way, enough cells become available for daily experiments and monthly papers. Also, performing experiments becomes routine because the cells live in a very well-established and familiar artificial system that is easily manipulated. Cell lines are particularly handy for the methods of molecular biology. The allure of cell lines was too great for cancer scientists and other researchers to resist. Ambitious investigators ignore human tumors because they have instead what they need most for success—an experimental system that allows a great amount of data to be produced easily.

Malignant cells living in a tumor, just removed from a patient, are much more difficult to work with. Various types of normal cells are often present in a tumor, and the tumor itself remains alive for only a short period of time. It is difficult, if not impossible, to measure accurately the effects of experimental procedures on only the cancerous cells while they are alive. These difficulties slow the work and limit the kinds of experiments and analyses that can be performed. As a result, fewer papers are published by scientists who study tumors than are published by those who study cell lines.

Almost all of the subdivisions of cancer research owe their existence to the facility of using cell lines as research tools. In addition to molecular biology, immunology, tumor cell biology, biochemistry, carcinogenesis (cancer initiation), endocrinology (the influence of hormones on cancer), drug development, and virology all rely on information derived from the cell line model.

Superficially, the use of cell lines in cancer research appears to be a logical solution to the difficulties of working with actual tumors. There is solid evidence that malignancies are of clonal origin—meaning that they are derived from one, original malignant cell by the process of cell division, or mitosis. Each cell is a copy of every other cell. The ancestry of a tumor cell line can be traced back through cells of the primary culture to the tumor cells living in an organ and finally to the first tumor cell within that organ. Studying a tumor cell line should, therefore, be very similar to studying the original tumor cells, only much easier, since the cells are living in a laboratory dish rather than in a person.

In theory, perhaps—but the realities of life in a petri dish are very different from the realities of life in the human body (see Chapter Five). The transfer from an *in vivo* to an *in vitro* environment precipitates a crisis period for cells. The one that survives a crisis period and learns how to live forever in culture undergoes fundamental changes that render it and its descendants profoundly different from cells that live in patients. The nature of life itself changes.

## THE PROBLEM

The environment in which a cell line lives and grows is the equivalent of a small saltwater fish tank without the fish. And the film of cells is similar in appearance to scum found on the bottom of fish tanks. These are artificial conditions, produced by humans and never found in nature. It therefore follows that cells that will thrive forever under such unnatural conditions must themselves be artificial, an artifact.

Cells—be they plant or animal—are remarkably adaptable. When environmental changes are sudden and dramatic, some cells successfully adapt by making radical transformations. When a cell adapts to permanent life on the bottom of a culture dish, it takes an evolutionary step that enables it to survive and thrive in its new environment. The cell line that results is neither human nor animal. It might just as well be from outer space.

The cancer establishment is so committed to cell lines that it disregards the characteristics of human cancer, which pathologists and others have documented repeatedly for more than 100 years. Instead, they ascribe the unique features of cell lines to human cancer, even though they are very different.

For example, when today's researchers discuss "lung cancer," they usually are not referring to lung cancer in the body but to cells derived from lung cancer that have lived for years in plastic dishes in a laboratory. The distinction between cell lines and human cells *in vivo* is no longer being made, largely because cell lines are so ideal for the methods of molecular biology.

The presidential address of a newly elected president of the American Association for Cancer Research (AACR) beautifully illustrates the field's devout faith in molecular biology as the great hope for cancer research. Speaking in 1989, at the beginning of his one-year term, Dr. Lawrence Loeb of the University of Washington School of Medicine praised molecular biology, stating that it "offers both the power and promise to unravel the mysteries of human cancer." Loeb and his colleagues believe that cancer research was previously limited by methodology and that "the absence of this limitation is the promise of molecular biology. I submit that molecular biology constitutes a series of new experimental approaches of unprecedented power—approaches that will increasingly change most aspects of cancer research."[3]

Although Loeb acknowledged "our lack of understanding of the molecular differences between normal and cancer cells" at the present time, he added, "I believe we are currently witnessing a quantum leap in our understanding of how cells work. This revolution is spearheaded by molecular biology."[4]

When these sentences were read, I immediately thought, "What cells is he talking about?" Going through many of the research papers referred to by Loeb in his address, including more than twelve from his own laboratory, I found my answer. Loeb and his colleagues depended heavily on cell lines, so these must be the cells that were yielding their secrets to the probings of molecular biologists.

Loeb and others in cancer research look to the experimental approaches of molecular biology to solve the cancer problem because of their power to produce information. The techniques of molecular biology enable scientists to break apart genes, analyze their DNA, and produce volumes of data. Because these techniques are so productive, experimental models are chosen not on the basis of their similarity to the natural situation but because of their convenience as research tools. Cell lines are the epitome of the convenient research model. But powerful methods used to probe a poor model of human cancer produce poor-quality information of little practical value.

Loeb's successor as president of the AACR in 1990 was Dr. Harris Busch, of the Baylor College of Medicine in Houston. Busch studies the HeLa line. Like Loeb, and like every other major researcher studying cell lines, Busch and his group believe that the behavior of the cells in their petri dishes is analogous to the behavior of cancer cells in the human body. And the rest of the cancer field follows the example of their elected leaders.

## THE SCHISM BETWEEN RESEARCH AND REALITY

Most cancer scientists are unaware of the fact that cell lines are an inappropriate model for human cancer. On the contrary, they truly believe that cell lines are good models for cancer research—because that is all they know. In graduate schools and research laboratories around the world, promising young scientists are told that cell lines are the standard for determining what cancer should be like. They never ask themselves if the behavior of cell lines mimics that of the real disease. Indeed, few of the molecular biologists involved in cancer research will ever see an actual human cancer.

Most pathologists perform practical tasks, such as identifying human cancers by analyzing cellular characteristics under a microscope. Cancer scientists have different interests. They make inferences about the nature of human cancer based on the behavior of cells that exist only on the bottom of a petri dish. As a result, there is little communication between surgical pathologists (who know a great deal about human cancer) and cancer researchers (who do not). The two groups undergo different training, use different technical jargon, usually work in different buildings, attend different meetings, and read different journals. The two sides of the cancer coin—clinical and research—exist in almost complete isolation from each other.

One of the primary clinical journals read by pathologists and oncologists is *Cancer*, which is published twice a month by the American Cancer Society. The papers in *Cancer* contain the most reliable information about the disease because the model observed by their authors is us—living, breathing humans. The papers published in *Cancer* reflect the realities of cancer as it occurs in the human body.

The leading experimental journal, however, is *Cancer Research*, published by the American Association for Cancer Research (AACR). Most of the papers published in *Cancer Research* are reports of cell line research. In issue after issue, I have found that more than half of the research papers concern cell lines. The number of papers that deal directly with any other cancer models, including human tumors, make up only a small percentage of the total research published. Data from cell lines are clearly the currency for scientific expression for the entire field.

Given the widely divergent models of cancer that are studied in the research papers featured in *Cancer* and *Cancer Research*, it is not surprising that the pictures they present of cancer are equally divergent. These contradictions between the real world of cancer in humans and the artificial world of cancer in a petri dish are at the heart of the cancer research scandal.

# 5

# The Contradictions of Stability and Differentiation

*A spontaneous eruption of carcinoma is succeeded by an identical eruption at another site.*[1]

Joseph Claude Anselme Recamier,
internist, 1829

*The character of the epithelium at the point of origin determines the form of all developing cancer cells, which then retain their once inherited character in all metastases.*[2]

Wilhelm Waldeyer,
pathologist, 1872

The contradictory knowledge of pathologists and most cancer researchers can best be understood by first considering the reality of cancer as it is seen by pathologists. To use an all-too-common example, consider the case of a woman who has undergone a modified radical mastectomy to remove a cancerous breast and adjacent lymph nodes. Although her physician believes all the malignancy has been removed, she continues to have regular examinations because it can take years for microscopic nests of metastatic tumor cells to grow large enough to be detected.

Four years after the surgery, the doctor's physical examination shows no problems, but a standard chest x-ray reveals a nodule about one-half inch in diameter in the right lung. This nodule was not present on a similar x-ray study taken six months earlier. Suspecting a recurrence of the woman's cancer, the physician immediately schedules the patient for a biopsy procedure. The biopsy specimen is obtained without difficulty and sent to the pathology laboratory for evaluation.

One day later, thin, stained sections of the biopsy are mounted on slides and ready for study. The pathologist has also retrieved from departmental files four-year-old slides of the breast tumor. The pathologist places a biopsy slide on the stage of a microscope, turns on the illumination, and looks into the eyepiece. The pathologist moves the slide around, noting the many groups of closely packed epithelial cells with large, intensely stained nuclei lying in a supporting framework of vascularized connective tissue. The other slides of the lung biopsy are also examined. After several minutes of careful observation, the pathologist decides it is cancer. Cancer cells are abnormal, and pathologists have been trained to recognize the abnormalities.

The pathologist then examines slides from the four-year-old breast tumor. After studying them carefully, the pathologist is sure that the lung malignancy is a metastasis from the original breast tumor, not a new tumor that originated in the lung or elsewhere. The microscopic structure, or histology, of the tumor found in the lung biopsy is identical to the histology of the four-year-old breast tumor.

Hospital pathologists, who deal with this situation almost daily, know that even though cancers enlarge, invade adjacent body parts, and travel to distant metastatic locations, they remain unchanged. The characteristics of human tumors, with rare exceptions, are fixed for the life of every tumor, regardless of where or when distant metastases of the tumor are found.

In 1874, Dr. W. Moxon, an English pathologist, described "rectum in liver," referring to rectal tumors that were growing in their original unchanged forms after metastasizing to the liver.[3] Scientists who study embryonic development would say that human tumors are *determined*. That is, their cells have been switched into a developmental path that cannot be changed, even when the malignant cells are growing in unfamiliar organs.

There are many examples of the constancy of human cancer. One is that tumor cells that produce certain products always do so, regardless of where they metastasize in the body. Colon tumors that produce a substance called carcinoembryonic antigen are always positive for it. Likewise, a prostate tumor that is diagnosed early because prostate specific antigen (PSA) was detected in the blood will continue to produce PSA years later at a metastatic site.

Another fixed feature of tumor cells is their level of development, or differentiation. Those that are poorly developed remain so, whereas well-differentiated tumors remain well differentiated. Tumors also behave in a consistent and fixed fashion. Those that grow rapidly and are likely to metastasize remain aggressive, whereas less aggressive cancers rarely change their behavior.

Fixed characteristics, of course, are also features of all normal cells. The last generation of cancer scientists understood the close relationship between normal and malignant cells. Dr. Eugene D. Day, writing in 1961 as a guest editor in *Cancer Research*, said that by studying natural tumors in animals, "those subtleties inherent in cancer which distinguish it from normal tissue may emerge clearly and simply."[4] Day, like pathologists the world over, knew that the differences that separate tumors from the normal tissues in which they arise are subtle.

## CANCER'S ESSENTIAL GENETIC STABILITY

Every nucleus of every cell in our bodies contains the same forty-six flexible, threadlike chromosomes, carrying identical genes. Despite the fact that every body cell contains the same genetic information, they do not all differentiate in the same manner because the *activity* of individual genes varies from body-cell type to body-cell type. The mix of genes that is active in each of the 200 or so different cell types of the human body is always different. For example, at any given moment, stomach cells and muscle cells will have different genes "turned on"— leading to the expression of the unique, differentiated features of their respective cell types.

These observable characteristics of a cell are called its *phenotype*, and the chromosomal characteristics are called its *karyotype*. Both normal and malignant human cells exhibit fixed phenotypes. Their observable characteristics do not evolve or change, indicating that their chromosomes and genes must remain stable as well. Both normal and malignant human cells, when living *in vivo*, are genetically stable.

It is true, however, that most human tumors do not have a normal karyotype of forty-six chromosomes. Malignant cells often have numerical chromosomal changes and structural chromosomal defects called mutations. But that abnormal karyotype also tends to remain stable over time. For example, if a breast tumor has a karyotype of eighty chromosomes, with several having mutations, metastatic cells from that tumor, discovered years later, will probably also exhibit eighty chromosomes and the same chromosomal mutations.

The scientific literature contains many studies that document the essential genetic stability of human tumors. One such study was conducted by Dr. Niels Atkin of Mount Vernon Hospital in Middlesex, England. Published in 1989 in the journal *Cancer Genetics and Cytogenetics,* the paper traced the progress in the 1980s toward an understanding of the genetics of human cancer.[5] Atkin reviewed eighty-one different papers and concluded that "the great majority of human tumors are monoclonal" and that each tumor "generally maintained its chromosomal characteristics."[6] In other words, not only do malignancies arise from one original malignant cell, but cells arising from this initial malignancy tend to be genetically stable, just like normal tissues.

## INSTABILITY AND PROGRESSION

In contrast to the known genetic stability of both normal and malignant human cells *in vivo,* genetic *in*stability is an inevitable side effect of a cell that survives a crisis period and emerges as an immortal tumor cell line from a primary culture of tumor cells. The number and structure of the chromosomes within immortalized cells in culture change randomly over a long period of time. For example, MCF-7 cells—the favorite experimental model of breast cancer—have been described as exhibiting "continuous chromosomal rearrangements and numerical changes."[7]

The pronounced genetic instability of the MCF-7 line was first described by a group at the National Cancer Institute in 1983.[8] However, good scientists that they are, these researchers did not suggest that this instability made MCF-7 a poor model for breast cancer. On the contrary, they also assigned the observed instability to human breast tumors. Attributing the genetic instability of MCF-7 to breast cancer is analogous to saying that since old cars rust, old people must rust as well.

Dr. Jorgen Fogh, as director of the Human Tumor Cell Laboratory at the Sloan-Kettering Institute for Cancer Research in New

York, had more than twenty years' experience working with tumor cell lines. In 1987, he wrote in the journal *Cancer Investigation* that "chromosome numbers vary from cell to cell in most human tumor cell lines."[9] Since the chromosomes contain the genes, cell lines are very genetically unstable because every cell of a line may have a different complement of chromosomes and genes.

This is the first contradiction between cell lines and the reality of cancer in humans. Pathologists know that the features of human tumors are fixed, whereas most researchers in the field believe in the concept of tumor progression. That is, they believe that the characteristics of human tumors are not fixed but change over time. This contradiction exists because cell lines, the favorite experimental model of cancer scientists, always progress and scientists assume that human tumors work the same way.

Every cellular characteristic that has ever been examined in culture evolves. Three characteristics that have been particularly well studied are growth rate, responsiveness to hormones, and tumor formation when cell lines are injected into animals. For example, a cell line that grows faster when estrogen is added to its culture medium might produce a subline that does not respond to estrogen. In another cell line, a few variant cells and their descendants might be able to form tumors when injected into mice, whereas the rest of the line cannot. Tumor progression has become the most important concept in cancer biology today because of the genetic instability that is characteristic of all cell lines.

The HeLa line, like MCF-7, is an archetypical human tumor cell line. It is probably the most studied of them all. For many years, it was the standard. However, as early as the 1950s, researchers knew that different clones of HeLa "exhibit differences in characteristics." They stated, "It is evident that the chromosome number in cells of the HeLa strain varies widely," and "changes in chromosomal constitution, especially in chromosomal number must be considered as mechanisms potentially responsible for the mutant phenotypes."[10] They understood that the HeLa's unstable phenotype was the result of its unstable karyotype.

At around the same time that the HeLa line was successfully established, other workers showed that normal, nonmalignant cells could also be established in culture, creating "normal" cell lines.[11] As the quotation marks around the word *normal* indicate, these cell lines are really misnamed.

In writing about the new cell lines (referred to in the 1950s as cell "strains"), these early investigators noted, "Evidence indicates clearly that changes in the characteristics of originally normal cell strains may also occur during long term culturing."[12] Like tumor cells, normal cells exhibit genetic and phenotypic instability when they are established in culture. Recognizing this fact, investigators in the 1950s were warning their colleagues away from cell lines. Needless to say, few researchers today heed the good advice of their predecessors.

It is hard to believe that all biomedical scientists are not aware of what happens to human chromosomes in cell lines. Consider a 1989 study in *Cancer Genetics and Cytogenetics*, which states, "After immortalization of human cells in culture, the previously very stable chromosomal complex has become unstable, spontaneously breaking and rejoining to form new genomic combinations."[13] The new chromosomal combinations included "rearrangements, duplications, deletions and trisomy." Trisomy means that instead of chromosome pairs, there is one or more triplets. The paper concludes, "Transformation of human cells to immortality is associated with gross chromosomal mutation changes."[14]

Today, the most popular "normal" cell line is the 3T3, which was derived from a ground-up rodent embryo. Although they are supposed to be models for normal cells in the body, these cells behave nothing like normal cells. They do not age, are unstable and undeveloped, have an abnormal number of chromosomes, exhibit chromosomal mutations, and are easily transformed to a malignantlike state.

The 3T3 cells have been in culture in laboratories throughout the world for a generation; yet because of their genetic instability, these cells no longer contain the normal karyotype of mouse cells from which they were derived. Instead, 3T3 cells exhibit the abnormal numbers of chromosomes and chromosomal mutations that are fundamental characteristics of malignancy. Normal cells in our bodies would not recognize 3T3 cells as even the most distant of relatives.

At the 1989 national meeting of the American Association for Cancer Research (AACR) in San Francisco, I enjoyed a talk given by one of the few pathologists I could find at the conference. Dr. Emmanuel Farber is the chairman of the Department of Pathology at the University of Toronto School of Medicine. He used the concept of progression to describe the changes that take place in normal liver cells before they become malignant. Surgical pathologists know that progression stops once a cell becomes malignant. Most cancer researchers do not. They believe that progression is a characteristic of malignant cells themselves.

After the talk, I introduced myself to Dr. Farber and told him that I enjoyed hearing someone who shared many of my views about cancer. In all the years of going to cancer conferences, it was refreshing to hear the truth for a change. I wondered if Farber fully realized the extent to which cell lines had corrupted the understanding of cancer that pathologists had taken 100 years to develop.

"Dr. Farber," I said, "I think it would be helpful when you use the word *progression* in a talk to define what you mean, because our understanding of the concept and what most believe is very different." At that point, I gestured toward the open door of the large meeting room at the Mosconi Convention Center to indicate the thousands of scientists at the conference. Farber looked in the same direction and, with a gesture of disdain, said, "They do not know anything."

The glamor of the gene-centered science of molecular biology has pushed pathologists such as Farber aside in the field of cancer research. The cancer industry—and the public—has become fascinated with such factors as genetic engineering, moving genes from one cell to another. The skills and knowledge of pathologists are no longer in vogue.

As a result, the cancer research community is almost devoid of people who understand human cancer. When the AACR was founded in 1907, six of its eleven charter members were pathologists. Eighty-five years later, most members of the association have never even turned the page of a pathology text. Molecular biology has pushed aside the very knowledge and expertise that make possible the diagnosis of all diseases.

The knowledge about cancer obtained from cell lines is routinely accepted as the truth because most cancer scientists have never learned the important features of human malignancy. In December 1990, when Dr. Bruce Alberts, a cancer researcher and professor at the University of California, San Francisco, admitted to the President's Cancer Panel that "we have rarely seen a real human tumor,"[15] he was speaking for a majority of the scientists in the field. Alberts's comments were made during a presentation before the President's Cancer Panel Quarterly Meeting in San Francisco in December 1990.

## THE QUESTION OF DIFFERENTIATION

Earlier in this chapter, I described the process by which a pathologist determined whether breast cancer had metastasized to a lung. The pathologist could make the diagnosis because both the original breast tumor and the tumor in the lung exhibited identical histologies—including ductlike structures that mimicked the features of normal, differentiated breast tissue.

Pathologists have known since the nineteenth century that tumors always retain features of their normal parent organs. When Moxon referred to rectal tumors that had metastasized to the liver as "rectum in the liver," he was saying not only that the characteristics of rectal tumors were fixed but that they related back to the rectum as well. The organ-specific differentiation of human tumors was made unmistakably clear to me in 1974, when Dr. Russell Sherwin and I demonstrated that breast tumors, just like the normal breast, produced secretions (see page 25).

By contrast, on the bottom of a dish, a lung tumor line looks like a breast tumor line, which looks like an ovarian tumor line, which looks like a prostate tumor line, etc., etc., ad infinitum. The differentiated features of the tumors are lost by the lines as they adapt to life under the artificial conditions of the culture environment. Cell lines show few of the signs of the differentiation that characterized their cellular ancestors because they have become undifferentiated (undeveloped) in culture.

It is also common for lines to lose their sex chromosomes, their two X or one X and one Y chromosomes, which indicate whether they were derived from a female or a male. These lines have lost their most basic differentiation, a sexual identity. No animal or human cell *in vivo* is without gender, but cell lines often are.

The differentiation of tumors, such as duct and gland formation and secretion production, is closely tied to how the malignant cells in the tumors behave and function. Cell lines, therefore, behave and function very differently because they lack differentiation.

This is the second important contradiction between the reality of human cancer and the beliefs of cell-line-based cancer research. The cells of human tumors always show evidence of the differentiated structure and function of the normal organs in which they arose, whereas cell lines do not.

The inability to distinguish one cell line from another because of the loss of cellular differentiation caused a scandal in cancer research in the 1970s. The details of this research debacle are chronicled in Michael Gold's book *A Conspiracy of Cells.*[16] The scandal involved the HeLa line, which is so hardy that if one HeLa cell is accidentally introduced into another culture, it will proliferate and choke out the other line. Sometime during the 1960s, just such contamination occurred. Breast tumor lines, prostate tumor lines, bladder tumor lines, and so on were contaminated and became HeLa. But since all cell lines look alike, none of the researchers using the contaminated cell lines and exchanging them with their colleagues knew what had happened. They did not realize that they were conducting experiments on cells that were actually descendants of cervical cancer cells removed from a doomed cancer patient in 1951. Cell lines are so undeveloped that a cervical cell can be mistaken for a prostate cell by experts.

By the mid-1970s, many laboratories around the world were studying HeLa cells without even knowing it. Millions of dollars were spent on projects concerning breast tumor cells, prostate tumor cells, and so on, when all the cells were really HeLa. Thousands of incorrect and misleading papers were written. It was not until Dr. Walter Nelson-Rees (of the Naval Biomedical Research Laboratory at the University of California's School of Public Health) became convinced that there were repeated instances of HeLa cells contaminating many other laboratory cultures in prominent scientific laboratories that the scandal was revealed.[17] If you cannot tell one cell from another, you have grave problems.

The exposure of the HeLa scandal may have brought the problem of contaminated cell lines to light, but the problem itself has not necessarily gone away. By now, HeLa or other cell lines may be everywhere, although being called something else, because it is impossible to distinguish one cell line from another.

As the quotations at the beginning of the chapter indicate, the basic characteristics of human tumors were recognized more than a century ago. Within the dynamic environment of the human body, both normal and malignant cells tend to exhibit stable karyotypes that maintain stable phenotypes. Tumors, like the normal cells in which they arise, are also always differentiated to some degree. Each has specific structural and functional characteristics that follow to some point the normal development of cells in the organ where it originated. The differences between normal and malignant cells are generally subtle.

In contrast, tumor and "normal" cell lines are very unstable. They exhibit gross karyotype instability that results in very unstable phenotypes. The act of culturing also leads to a loss of differentiation, and cell lines become undifferentiated.

It seems reasonable to conclude that an evolutionary-type step is taken by the occasional normal or malignant body cell that adapts to immortal life in culture. This step transforms a cell from one that is stable and differentiated to one that is not. Yet normal or diseased animal or human cells *in vivo* are not like that—which makes cell lines a new form of life on earth.

Evidence of these contradictions has been in the scientific literature for more than thirty years, and it has been systematically ignored by the cancer establishment. This is intellectual dishonesty of the highest order. It is a very large, growing scandal that our understanding of a disease that kills more than one-half million Americans every year is increasingly dominated by less than useless information. The ivory tower world of cancer is out of touch with reality, wasting our time and money.

Most of us picture cancer as a mass of wildly abnormal, undeveloped cells because that is what cell lines are. And cell lines are the model that dominates the field of cancer research. But this model has lost all fundamental links to the human disease. It has lost the differentiated features and stability of all human tumors. The cancer industry is being misled by its favorite experimental model and just passes along the incorrect information. Remember how the sun once revolved around the earth? It still does, everywhere men and women quest for knowledge about cancer. Everything is wrong because the models are wrong—the geocentric model of the solar system and the cell line model of the human cancer cell.

# 6

---

# The Contradictions of Initiation and Metastasis

*In every organ the carcinomas predominantly go through the same metamorphoses as those most frequently seen in epithelial cells under normal circumstances at the same sites.[1]*

Wilhelm Waldeyer,
pathologist, 1867

*. . . Differentiation and its dysfunctions . . . may well lie at the heart of the neoplastic transformation.[2]*

Sidney Weinhouse, Ph.D.,
Tufts University School of Medicine, 1972

For the public, the two most pressing questions about cancer are "Why does it start?" and "Why does it spread?" These are also the most pressing questions for cancer researchers, although they phrase their inquiries in somewhat different terms. In laboratories around the world, scientists regularly ask,

"How is cancer initiated?" and "How does it metastasize?" Not surprisingly, the answers differ, depending on the laboratory models the investigators use.

## INITIATION

Most of the body's organs contain a subpopulation of undeveloped, immature cells called stem cells. During life, these cells divide into two "daughter" cells, and it seems that only one of them goes on to develop into a replacement for a worn-out, fully developed, old cell that has lost the ability to divide. The other "daughter" cell is thought to remain a stem cell. This cycle of cell replacement occurs regularly at different rates in various organs.[3] In the intestines and bone marrow, for example, cell replacement occurs frequently because these cells have a lifetime of only a few days.

The differentiation that is characteristic of all cancers occurs because each tumor cell "remembers" the organ in which it was born and develops along the route that the organ's normal cells take during the cell replacement process. But in cancer cells, something goes wrong.

In fact, dysfunctions in cell developmental processes seem to be right at the center of the cancer problem. Cells are most sensitive to the effects of carcinogens during their developmental period. For example, mammary tumors of rodents are easiest to produce during adolescence, when mammary cells are maturing. Conversely, cells become less susceptible to cancer as they age and become more differentiated, apparently because they have lost the ability to divide. One-day-old developing intestinal cells divide and can become malignant, whereas three-day-old mature cells are not capable of dividing or becoming malignant. In tissues that do not undergo significant cell replacement—such as heart muscle and the nerves of the central nervous system—cancer is extremely rare.

These observations—gathered over more than 100 years of clinical and experimental research, as the quotations at the beginning of the chapter indicate—strongly suggest that cancer begins in differentiating (developing) cells. Maturing cells of the human body are the cancer's target. One of the best experimental studies on the topic appeared in the cancer literature in 1975. The paper—on cancer initiation in the rat—was written by Dr. Kazymir Pozharisski, a Russian pathologist from the Petrov Institute of Oncology in St. Petersburg.[4] Pozharisski's group injected 800 laboratory rats with a known carcinogen once a week for eight months. The animals were examined at different times over five months, and the development of intestinal tumors was followed under the microscope.

The intestine, like many organs of the body, has developmental zones where stem cells are located. These zones are several cell layers wide, with the stem cells at one end. Immature cells enter the zone at this end and, after several divisions, exit at the other as fully differentiated intestinal epithelial cells.

Pozharisski's experimental data suggest that the carcinogen affected developing cells in developmental zones and that intestinal cancer was produced in the rats because of "differentiation disorders" of these cells. Tumors "originate from cells that have lost the ability to differentiate."[5] That is, the impact of the carcinogen caused a cell to become frozen at an incomplete stage of development, depriving it of the ability to complete its differentiation. From that time on, the malignant cell was stable, exhibiting features of the developmental stage at which it was stuck, while retaining the capacity to divide because it was not fully mature.

The investigation of human lung cancer that I conducted in 1980 yielded similar findings. My conclusion, that cancer initiation alters the normal maturation of cells, was really the same as Pozharisski's.

In both studies, cells that were struck by malignancy early in their development appeared to be blocked there, giving rise to poorly differentiated tumors, whereas those struck later in their development became well-differentiated tumors. The results of both sets of experiments suggest that malignant cells are those that cannot complete developmental processes in organs that furnish replacements for dying, old cells. Instead, malignant cells seem to be arrested at some point during their development and stop differentiating and aging.

Few cancer researchers in the United States today would consider a project like Pozharisski's. The experiments alone took five months to complete, and the data from 800 animals probably took at least another year to analyze. How could a researcher publish enough papers to become successful and well funded if he or she spent so much time on a single project? Instead, bright young (and not-so-young) cancer researchers are diligently studying the process of cancer initiation in "normal" cell lines, producing reams of research papers in the same time that Pozharisski and his associates devoted to a single study.

For some, the diligence has paid off. In 1989, Dr. J. Michael Bishop and Dr. Harold Varmus of the University of California, San Francisco, were awarded the Nobel Prize in Medicine for their work on cancer initiation. Bishop and Varmus discovered that normal cells contain genes that are believed to cause cancer when they malfunction (see Chapter Seven). This "oncogene" theory of cancer initiation has been heralded as the future of cancer research—the key that will unlock the riddle of human cancer. Unfortunately, the experimental work that led to the Nobel Prize did not initiate cancer in normal cells with "oncogenes"—researchers used a "normal" cell line, 3T3.

As I noted in the last chapter, 3T3 cells have been in culture for a generation. Like all cell lines, these cells have an abnormal number of chromosomes and chromosomal mutations, basic characteristics of malignant cells. As a matter of fact, "normal" 3T3 cells even form tumors when inoculated into some kinds of mice. In addition to exhibiting genetic and phenotypic instability and being immortal, the 3T3 line is very close to, if not actually, malignant.

In conducting their research, Bishop and Varmus and others used 3T3 cells to test the transforming activity of the supposed oncogenes—inserting the "oncogenes" from human tumors into 3T3 cells and watching for cellular changes such as a piling up of cells.[6] As we will see in the next chapter, the results of these experiments are unreliable and far removed from human cancer.

The extent to which Bishop and Varmus relied (and undoubtedly continue to rely) on cell lines can be inferred from the references cited in a major point-of-view paper published by Bishop in *Science*.[7] The paper contains more than 300 references to research papers of other scientists. In about 70 percent of them, the experimental model studied was cell lines. Thus, about 70 percent of the information in Bishop's paper was about cell lines, not human cancer. Like most of his colleagues in the field, Bishop knows a great deal about "cancer initiation" in cell lines, and like most of his colleagues, he probably believes that this knowledge has a close relationship to cancer initiation in humans.

Not all molecular biologists share this misguided faith in cell lines. Dr. Peter Duesberg, a noted molecular biologist at the University of California, Berkeley, concluded in a 1987 paper in the *Proceedings of the National Academy of Sciences* that "normal" cell lines are not reliable in tests of the cancer-causing activity of genes.[8] How could they be, when cancer strikes stable, normal body cells during their development and aging, whereas the cell lines used to represent them are unstable, ageless, and undeveloped?

Cancer of the heart muscle or nerves of the central nervous system is very rare in adults because these organs do not have a supply of stem cells. Without stem cell divisions, there are no newly formed heart or nerve cells that can be struck by cancer. (Unfortunately, the lack of stem cells also explains why a damaged heart or severed spinal cord cannot regenerate.) Only differentiating, aging cells are susceptible to cancer. Researchers, including Nobel laureates, must understand this fact before they attempt to study carcinogenesis in cell lines. Data from a model that neither develops nor ages cannot be relevant to the cancer process in humans.

During the 1988 meeting of the AACR in New Orleans, I discussed the connection between cancer initiation and cell differentiation with a young postdoctoral fellow from the NCI. Not surprisingly, his work involved inserting genes into 3T3 cells. After our conversation, he seemed to understand that everyone has been misled by data from 3T3 cells. I would not bet, however, that he stopped using them as normal cells when he went back to his laboratory at the NCI. The vested interests of the cell line bandwagon in biomedical research are enormous.

## METASTASIS

The most perplexing problem for oncologists—metastasis—actually comes after a cancer has been initiated. Cancer patients rarely die from a primary tumor. It is the metastases—invading and destroying vital organs like the lungs and the liver—that kill most cancer patients. When science discovers the mechanisms that permit a tumor cell to detach from a primary tumor and "colonize" other sites in the body, it will have taken a giant step toward the control of cancer. Effective treatment will come when the metastatic process can be stopped. As a result, studies of metastasis have always been a high priority for cancer researchers.

Tumor cell lines began to be used by researchers in experimental metastasis studies in the 1970s.[9] In 1988, the earliest proponent of such work—Dr. Isaiah Fidler of the University of Texas, M.D. Anderson Cancer Center in Houston—was given the prestigious G.H.A. Clowes Memorial Award by the AACR for his research achievements. By 1989, experimental metastasis studies were so popular that 500 scientists from around the world attended a meeting devoted to this one topic.

The popularity of cell lines as a human cancer model in metastasis studies is not just due to the fact that cell lines grow well in culture; they also grow well in animals. In a typical experiment, cells of a line are injected intravenously or implanted under the skin of mice. Three weeks later, after the cells have spread through the blood circulation, the animals are sacrificed and the nests of cells growing in various organs are collected and studied. Thousands of studies have provided a very clear picture of the metastatic process—as it occurs in tumor cell lines.

Not surprisingly (given the documented instability of cell lines), the picture is one of pronounced change and instability. Because of this research, cancer doctrine now states that cells with high metastatic potential are rare in tumors. They are supposed to arise only on the rare occasions when a particular but unknown characteristic is suddenly "turned on," giving this rare cell the ability to metastasize.

Compare this picture with that provided by thousands of observations of countless human tumors. Pathologists around the world know that the risk of metastasis is inversely related to the level of tumor differentiation. Poorly differentiated tumors are generally aggressive, with high metastatic potential, whereas well-differentiated tumors are usually less aggressive, with lower metastatic potential. These levels of differentiation—and corresponding metastatic risk—are fixed for the life of the malignancy. High metastatic potential seems to be a stable characteristic of many (if not all) malignant cells of a poorly differentiated tumor. Yet the experts tell us that high metastatic potential is not fixed and not a characteristic of tumors but of only a few variant cells within them. According to the experts, all tumors start out with a low potential for metastasis. Metastases, when they do form, are the result of a few cells that have evolved into highly malignant forms.

This misguided theory of metastasis has become dogma because experts routinely assign the qualities seen in cell lines to human cancer, without investigating the reliability of the information. Instead, scientists proclaim that "most of the pieces of the cancer puzzle are now in place" and that "success is just around the corner." They make these pronouncements in scores of review articles, with titles such as "Genetic Instability of Cancer, or Why a Metastatic Tumor Is Unstable and a Benign Tumor Is Stable."

A 1981 paper on metastasis, "Metastatic Potential Is Positively Correlated with Cell Surface Sialylation of Cultured Murine Tumor Cell Lines," is illustrative of the misinformation that confronted me when I attempted to discuss my own research findings on metastasis.[10] These researchers said that highly metastatic mouse tumor cell lines have more surface sialic acid than poorly metastatic lines—a finding that directly contradicted my own findings in human cancer, published seven years before (and discussed in Chapter Three).

The phenomenal inaccuracy of the information about metastasis that has been acquired from cell lines is further illustrated by information from human autopsies. Those involved in pathology (and in cancer surgery and treatment) have long been aware of the fact that most cancers metastasize to the lungs or liver. In 1963, a study published in the journal *Cancer* compared the distribution of primary tumors and metastases in 317 autopsied patients.[11] All the common forms of cancer were represented. The data show that 90 percent of the tumors spread either to the lungs or to the liver or to both. Human tumors metastasize primarily to these two organs because of their types of blood supply and central anatomic locations.

Cell line researchers, in contrast, come up with markedly different findings when they investigate metastatic distribution. When tumor cell lines are injected into the veins of laboratory animals, metastatic lesions appear virtually everywhere, in many different organs.

Once again, the data from cell lines contradict what is known of the human disease. But this unstable model dominates experimental metastasis studies. The information derived from this work is less than useless because it is better to believe nothing than it is to believe something that is misleading and incorrect. This false model leads researchers deeper into the darkness, away from the truth of—and perhaps the cure for—cancer.

# 7

# The Elusive Oncogene

*There is still no proof that activated proto-oncogenes are sufficient or even necessary to cause cancer.*[1]

Peter H. Duesberg, Ph.D.,
Department of Molecular and Cell Biology,
University of California, Berkeley, 1985

The most popular, well-funded, and establishment-serving belief about cancer to be supported by cell line theory may be the much-heralded "oncogene" theory of cancer initiation. This theory, which has already garnered one Nobel Prize and countless research dollars, is the central belief of modern cancer doctrine. In a few short years, this astonishingly weak explanation of cancer's earliest stages is widely accepted as the final word on the subject. But acceptance is not synonymous with truth.

## BASIC GENETICS

Most people in the cancer field now believe that cancer is the result of mutations, or structural and chemical alterations of genes. The initiating event is thought to strike and alter a gene in a chromosome within the nucleus of a normal cell.

The entire genetic code of an individual is contained within the genes, strung like beads along the twenty-three pairs of chromosomes found in the nucleus of every human cell. (The exceptions are the reproductive cells—sperm and ova—which have only twenty-three chromosomes, one from each pair.) Coiled inside each chromosome is a several-foot long, double-strand molecule of DNA. Genes are various-length segments of chromosomal DNA.

Each gene is a template, or guide, for the production of a specific protein. Within the cell, genes issue the instructions by which cells build proteins from a pool of twenty specific amino acids. These instructions are transferred from the cell's nucleus to the protein "construction site" in the cytoplasm by a single-strand copy of DNA known as messenger RNA.

Proteins govern how our bodies look and function, from eye color to the strength of our muscles. Life is possible because some proteins function as enzymes or catalysts, which make possible the thousands of chemical reactions that take place continuously within cells. Without them, the rates of these reactions would be too slow for life to be possible.

When genes are mutated, proteins can become defective and cells can indeed become sick. But do they become malignant? Is the ultimate cancer target within a cell really a gene? Does its mutation turn a cell malignant?

Molecular biologists have convinced almost everyone that the answer is yes—cancer reduces to alterations of DNA. They present this theory as if it were an unequivocal truth about nature inscribed in granite. Definitive statements in the scientific literature, such as "cancer results from mutations," are commonplace.[2] After all, cancer research is the realm of molecular biologists; molecular biologists study genes; therefore, altered genes must be responsible for cancer. Case closed.

Few people seem to remember that less than a quarter century ago viruses, not genes, were thought to be responsible for cancer. The evolution of the viral theory of cancer into the oncogene theory of cancer is a classic example of how the cancer establishment manages to preserve its image even in the face of evidence that refutes its doctrine and favorite models.

## THE VIRAL THEORY AND ITS DESCENDANTS

In 1969, two virologists at the National Cancer Institute proposed a new version of the viral theory of cancer, which had been in existence for years.[3] They said that cells contained viruses hidden in genes, which when activated, caused cancer. Hundreds of millions of dollars were spent on the search for these hidden viruses in human cells during the next decade, with no success. But the experts could not admit that they were wrong and that all those millions of research dollars had been spent on a wild goose chase. Instead of abandoning the viral theory, the theory was dramatically altered. It is not viruses within genes that are the cause of cancer, said the experts, it is the genes themselves!

The oncogene theory proposes that responsibility for cancer depends not on hidden viruses in genes but on about twenty specific genes called *proto-oncogenes.* It is believed that hundreds of millions of years ago, certain genes that were unique to living cells were incorporated into several different types of viruses when the viruses infected cells. The theory says that each of these cellular genes eventually evolved into a "viral oncogene," which can initiate cancer when certain animals are infected by the virus. The same genes, remaining behind in the cells of our very early ancestors, evolved in a different fashion. They have now, as always, normal functions in our cells and are called the proto-oncogenes.

Proto-oncogenes are supposed to be normal parent forms of oncogenes. They have functions within cells, as do all genes, and do not cause cancer. However, when mutated, proto-oncogenes are believed to malfunction and become oncogenes that can cause cells to become malignant.

The oncogene theory has been heavily promoted by virologists (molecular biologists who study the genetic material of viruses) who did not want to give up their influence within the cancer research establishment. Even though a great deal of work and money, over many years, have shown that viruses do not cause the vast majority of human cancers, this has not stopped virologists from keeping a very large slice of the research pie.

As was discussed in Chapter Six, the supposed oncogenes from human tumors are often inserted into the 3T3 "normal" cell line to test their ability to change several cellular characteristics. The 3T3 cells are said to be transformed to a cancerlike state when they become rounded and pile up in culture.

But 3T3 cells do not need an additional gene to be considered malignant. These cells are so susceptible to transformation that simply changing the concentration of nutrients in a culture medium will prompt transformation. If you so much as look at 3T3 cells disapprovingly, they will become rounded and pile up.

Cancer tests using 3T3 cells are also flawed because of the testing methods employed. Suspected oncogenes are introduced into 3T3 or other cells by physical means or by certain viruses, which usually results in the establishment of multiple copies of the gene in the chromosomes of the recipient cells. As many as 100 copies may be inserted at many different chromosomal sites.

Multiple copies of a gene, integrated into host chromosomes at numerous places, are very different from a single copy integrated into a single chromosomal location. Yet molecular biologists rarely if ever note in their research papers how many copies of a supposed oncogene are present or where they are located in the chromosomes of cells that have acquired new genetic material.

Molecular biologists also avoid discussing the fact that the procedures for transferring genetic material themselves produce mutations of genes.[4] Genes that are being inserted into cells often become structurally altered and are found to contain deletions, duplications, and mutations after insertion is complete. As a result, researchers can never be sure if the gene with which they began an experiment is the same once it gets inside the recipient cells.

## THE FAILURE OF THE ONCOGENE THEORY

Perhaps the most damning evidence against the oncogene theory is the fact that supposed human oncogenes do not transform true normal cells, which have a normal set of chromosomes. Furthermore, there is absolutely no evidence from observations of human tumors to indicate that the mutation of any proto-oncogene is essential for cancer initiation. In fact, in many tumors, all the supposed proto-oncogenes are normal; there are no oncogenes present.

Two conversations I had while at the 1990 meeting of the AACR in Washington, D.C., illustrate the shaky ground on which the oncogene theory rests. The first took place at what is called a poster session of papers dealing with the relationship between genes and cancer. A research group from an Ivy League medical school had conducted a study on pancreatic cancer—a particularly deadly form of cancer that is allegedly caused by a mutation in a proto-oncogene known as the *ras gene*.

The researchers explored the relationship between the ras gene and pancreatic cancer by using an animal model of the disease process. Animal cancer is often a good model of human cancer. Pancreatic cancer was induced in rats by a chemical carcinogen. The resultant tumors were then examined for evidence of the ras mutation. As one of the authors of the study explained to me, "I asked the question, is our rat model for pancreatic cancer mutated in the ras gene? It is an interesting question. But the rat pancreatic tumors do not have the mutation."

In other words, the project failed to show the expected association between rat pancreatic cancer and the ras gene. The researcher had to consider the possibility that the ras gene mutations found in many human pancreatic tumors were not responsible for the disease. The mutations could be coincidental or secondary to the cancer-initiating event. In a single brief conversation, a cancer scientist admitted that his data did not support the oncogene theory.

However, this group has not given up trying to associate the ras gene with rat pancreatic cancer because that is dogma. Scientists cannot become prominent by publishing data that do not support the oncogene theory. Publications on rat pancreatic cancer coming from this laboratory would be ignored by most in the field if they contained information that did not fit in with established beliefs about cancer. Therefore, the group is now looking at the activity of ras in tumors. The ras gene controls the production of a protein called p21; the researchers reason that since they found no gene mutation, perhaps the tumor cells make more p21 than normal pancreatic cells. Yet even if overproduction of p21 were found in rat pancreatic tumors, the significance would be far from clear. The protein p21 has been found to be associated with cell differentiation, and it is hard to imagine how more p21 in a cell would make it malignant.

The second conversation occurred at another poster session, with a scientist who had conducted research on the effects of dietary fiber on proto-oncogene expression in the rat colon. Like many researchers, his group had chosen a topic that was very trendy—and therefore fundable—at the time they were writing their grant. Research scientists can be very creative when finding money for research and quite agile as they jump on bandwagons.

For years, television and print advertising have admonished us to eat more fiber, extolling the potentially protective effect of fiber against cancer of the colon. These researchers, knowing that fiber was a popular topic in the media, struck while the iron was hot and received research funds from a company that made breakfast cereal.

"Oncogenes" have frequently, but not always, been associated with human colon cancer. In this project, the researchers fed rats test diets with different amounts of wheat bran to see if any proto-oncogene activity in the colon was changed by any of the diets. They hypothesized that high-fiber diets would reduce the level of the proteins coded by the proto-oncogenes in the rat colon.

As in the research on pancreatic cancer, the scientists were disappointed. "We were hoping to find that the more fiber rats ate, the lower the activities of the proto-oncogenes in the intestines, [in order to] get more funding from [food] companies," said the researcher manning the poster. "But it did not work out that way." Even in rat colon tumors, the activities of the proto-oncogenes did not change. Once again, the oncogene theory failed to predict observations of malignant cells in their natural environment.

In sum, many predictions based on the oncogene theory of cancer initiation fail to be upheld by empirical evidence. Mutations of proto-oncogenes or increased amounts of their proteins within cells are not required for the human disease. Cancer often occurs without them. However, instead of abandoning the oncogene theory, today's cancer researchers, like astronomers of the past, have simply made their theories more complicated in order to explain the failures of their model.

## DAMAGE CONTROL: REVISING THE ONCOGENE THEORY

When a model of nature is poor, the theories that are derived from that model inevitably become increasingly complex. Younger scientists, as they become heirs to established doctrine, must develop new and more intricate variations on the theories to make them fit what is actually observed in the real world. Consider Ptolemy's system of astronomy. According to Ptolemy, the planets moved around the earth in compounded circles. Predictions made with Ptolemy's system never quite conformed with the best available observations of the planets. For 1,500 years, astronomers tried to eliminate the discrepancies by making adjustments to Ptolemy's model of the solar system. As time passed, astronomy became so convoluted that a sixteenth-century astronomer said that no system as cumbersome as the Ptolemaic had become could possibly be true of nature.

Now consider the oncogene theory of cancer initiation. It began by proposing that twenty normal genes, the proto-oncogenes, were responsible for cancer. One mutation of one proto-oncogene within one cell was believed to be enough to cause the cell to become malignant. After a while the experts realized that "oncogenes" alone were not sufficient to explain cancer, so the theory was revised.

To explain the lack of a good correlation between the supposed oncogenes and human cancer, the concept of antioncogenes, or tumor suppressor genes, was invented. If mutated proto-oncogenes cannot be found in tumors, then perhaps there are mutated antioncogenes that act in a manner opposite to oncogenes. In their normal state, these antioncogenes are believed to suppress the development of cancer. When these suppressor genes are struck by mutations, the experts say that cells are more easily converted to malignancy.

And how are researchers going to find these alleged tumor suppressor genes? You guessed it—by inserting the suspected suppressor genes into unstable, undifferentiated tumor cell lines and watching for changes.

Today, it seems that every month a journal article announces the discovery of a new putative tumor suppressor gene. It is now believed that not 1 but 10 or more mutations of proto-oncogenes and tumor suppressor genes may have to accumulate in a cell before it becomes malignant. Furthermore, a committee of authorities recently identified an additional 179 changes of chromosomes in human cancer cells, which according to theory could be sites of additional genes involved in cancer initiation.[5] In breast cancer alone, one or more mutations in 13 of the 23 pairs of human chromosomes have already been described. Even as illustrious a member of the cancer establishment as Dr. Lawrence Loeb, a past president of the AACR, has expressed concern about the steady progression to greater complexity of the oncogene theory. "The dilemma is that there are too many mutations in human cancer," said Loeb in a 1991 paper.[6] This is probably only a fraction of how complex the oncogene theory will eventually become.

There may be as-yet-undescribed changes in genes and chromosomes that are associated with cancer. Scientists have sequenced only small portions of the DNA in tumors, and in a few years it is likely that hundreds, if not thousands, of gene and chromosome changes will be cataloged in human cancer cells. But then, as now, the experts will not know what the mutations mean to the development of cancer.

The one explanation that the experts will not consider, because it would invalidate a generation of research, is that the mutations are actually irrelevant to cancer initiation. The more mutations the experts discover, the more likely it becomes that all of them do not have the same importance. If any of the mutations are the *result* of cancer rather than its cause, the rest become suspect as well.

If the oncogene theory were really correct and cancer were caused by mutated genes, the germ cells—sperm and ova—should also contain these malfunctioning genes; and consequently, children should inherit cancer from their parents on a fairly frequent basis. Only diseases that are caused by mutations within the germ cells are passed to succeeding generations. One such genetic disease is cystic fibrosis, which appears (as most genetic diseases do) early in life. Chromosomal and gene abnormalities that occur in body cells during the life cycle are not inherited.

Yet parents do not pass cancer to their offspring as they do cystic fibrosis. Children with a high risk of getting cancer are extremely rare in human populations, much rarer than the 1 in 2,500 newborns who develop cystic fibrosis. Furthermore, the cancers of childhood that can be inherited and are probably initiated by gene mutations—such as retinoblastoma, Wilms tumor, and soft-tissue sarcomas—are hardly ever seen in adults. These facts suggest that exceedingly uncommon child-hood cancer of genetic origin is very different from the ex-tremely common form that appears much later in life as a consequence of aging. Thus, facts about malignancies in chil-dren point away from genetic to nongenetic events as the primary cancer-initiating events in adults. (The theoretical im-plications of this factor will be discussed in the next chapter.)

There is, however, an increased risk for certain cancers in some families. For example, it is known that the daughters and sisters of breast cancer patients have twice the risk of the general population for contracting the disease. Obviously a modest predisposition for breast and other cancers can be in-herited—but not the disease itself. A similar situation charac-terizes heart disease, in which high levels of blood cholesterol found in certain families increase the risk of the disease. In breast cancer, high levels of a circulating female steroid hor-mone have been reported in the daughters of breast cancer patients. These increased hormone levels could increase risk by stimulating breast cell division and increasing the number of cancer targets—developing breast cells. This sort of empirical evidence shows that the relationship of genes to most cancers is indirect, at best.

Given that not one "oncogene" or mutated "tumor suppres-sor gene" has been found to be essential for the initiation of cancer, what do these genes actually do in a cell? Like others, they contain genetic information for the production of specific cytoplasmic and nuclear proteins. Human studies—including one published in a 1987 issue of the *Proceedings of the National Academy of Sciences*—indicate that the protein product of at least one proto-oncogene is involved in the normal development of cells.[7]

This is an interesting but not surprising observation since hundreds of different proteins, if not more, are involved in differentiation. The observation reminds us that good representations of normal cells—in which developmental and aging pathways are functioning—should be studied to discover the role of mutated genes in cancer initiation. "Normal" cell lines such as 3T3 cannot provide correct answers.

For years, cancer experts have been arguing that 2 + 2 = 5. Volumes of misinformation about cancer have been created, validated, and legitimized by a bureaucracy intent on maintaining its credibility, legitimacy, and power. This misinformation will continue to be propagated for as long as cancer researchers serve the goals of a bureaucracy rather than those of science.

# 8

---

# The Epigenetic Theory

*It is important that in studying the mechanism of carcinogenesis we must look at the early events and keep our minds open to the possibility that the primary target is not DNA.*[1]

I. Bernard Weinstein, M.D.,
Columbia University College of
Physicians and Surgeons, 1968

In the course of almost 100 years of cancer research, a number of theories about the cause of cancer have come and gone. Some, such as the viral theory, have simply evolved into new variations on old themes. Others were left to languish in obscurity because they were incompatible with prevailing scientific opinion. One theory has been influential for decades and deserves serious consideration today.

## WHICH CAME FIRST?

In 1902, a German scientist named Theodore Boveri proposed a somatic mutation theory of cancer causation. Boveri hypothesized that cancer was the result of chromosomal abnormalities within body cells. In many respects, the oncogene theory of cancer is just a new wrinkle on Boveri's idea that cancer has a chromosomal (genetic) origin.

Although "proto-oncogene" and "tumor suppressor" gene mutations are not consistent markers of human cancer cells, changes in chromosome number and structure usually are. There is no question that the chromosomes within a cell's nucleus are affected by cancer, but the cytoplasm of a cell also contains markers of human malignancy. These cytoplasmic markers may be significant clues to the nature of the earliest events in carcinogenesis.

In normal cells, each advancing stage of cellular development features different cytoplasmic structures—called organelles—that play particular roles in the functioning of the cell. In developmentally blocked cancer cells, these organelles are also blocked and do not mature. In the lung tumors I studied, malignant cells appeared frozen at stages of development in which their cytoplasm lacked ciliary components. Without all the parts, cilia could not be assembled.

Well-differentiated cancers, which closely resemble normal tissues, tend to have only minor chromosomal abnormalities. Poorly developed tumors, in contrast, usually have wildly abnormal chromosomes. There is a reciprocal, mutually dependent relationship between genetic (nuclear) alterations and developmental (cytoplasmic) alterations in malignant cells.

These observations lead to a basic dilemma. Which alteration comes first? Do the chromosomal abnormalities in the nucleus come first, and then block cytoplasmic development? Or is the reverse the case? This "chicken or egg" question is the most important consideration for a correct understanding of cancer initiation. If research efforts are focused on the wrong site, researchers will end up studying only secondary changes and never discover the ultimate molecular targets or fundamental defects that render cells malignant.

For many years, cancer scientists have looked to the genetic material in the nucleus for the answer to the cancer riddle— generating theories with many shortcomings. The evidence derived from cell lines in support of the oncogene theory is unreliable and unconvincing. Moreover, observations of the behavior of cell lines when they are returned to an *in vivo* environment seem to support a markedly different interpretation of cancer. Perhaps it is time to look somewhere else.

Human tumor cell lines are undeveloped because they have adapted to immortal life in culture. This lack of development changes dramatically, however, when the lines are inserted into animals.[2] The tumors that grow in the animals exhibit a degree of development similar to that of the original human tumors from which the lines were derived so many years ago. The cell lines' loss of differentiation is reversed when they are returned to the *in vivo* environment. This *in vitro/in vivo* reversibility of differentiation is particularly true early in the evolution of cell lines, before they have accumulated a large number of additional chromosomal abnormalities because of life in culture.

When undifferentiated and unstable cells from a cell line are injected into an animal, the return of cytoplasmic development is also accompanied by chromosomal stability. It appears that the stability of chromosomes within cells is dependent on the acquisition by the cells' cytoplasm of a degree of differentiation that is only possible in living animals.

These findings suggest that the ultimate control over both differentiation and chromosomal stability within a cell resides in signals coming from other cells in the body. Some of these signals are chemical, others are electrical, and still others are mechanical. They are received first by the cell membrane, are modified in the cytoplasm, and finally are sent to the nucleus in the form of regulatory proteins. It is these regulatory proteins that control cell development, by turning genes on and off in ways we do not yet fully understand.

## POSSIBLE EXPLANATIONS

When a cell's nucleus is suddenly cut off from the signals sent by the body's other cells, it is conceivable that genes will not be turned on as they should be and that the cell's chromosomes will lose their stability, perhaps acquiring the mutations associated with cancer. Without the cues from the rest of the body, the cell's development could be blocked and as a result might be transformed to a malignancy.

In this view, the alterations that render a cell and its off-spring permanently malignant are not genetic but *epigenetic.* That is, the loss of gene activity is responsible for the initiation of cancer, rather than changes in the chromosomes or the genes themselves, which are secondary events. Cancer is the result of a breakdown in the signaling mechanisms that control the orderly activity of genes in developing cells.

The extent of damage to normal cellular development is clearly variable. If many genes are silent early, large deviations from normal development would result. Under the microscope a pathologist would see poorly differentiated tumors frozen at early stages of development. At the other end of the spectrum are well-differentiated tumors and minimal-deviation leukemias, in which genes are silenced late in development, so the malignant cells retain the capacity for quasi-normal development.

In normal cells, aging is characterized by several chromosomal changes that are strikingly reminiscent of the changes in malignant cells. Among them are numerical changes of chromosomes, abnormal chromosomal structures, alterations of methylation patterns of genes, and altered telomeres.[3] One of the building blocks of DNA is sometimes modified by the addition of a methyl group ($CH_3$). Genes are turned off when methyl groups are attached to this building block within their regulatory regions. Also, chromosomes have specialized caps at their ends called telomeres, which are shortened as cells age.

The normal chromosomal changes of cellular aging, just mentioned, are also characteristic of malignancy and could have a common origin—the lack of reception by cell nuclei of signals sent by other body cells. Old cells, like old people, may have deteriorated senses, and malignant cells may have damaged sensory mechanisms.

The presence of similar chromosomal alterations in malignant cells and aging normal cells suggests that such chromosomal changes are not necessarily the cause of the disease but a coincidental or secondary development. The altered chromosomes—and sometimes the altered genes—of malignant cells might be the result of developmental and aging operations within the cells as they try to complete their blocked developmental programs. For example, a loss of telomeres could destabilize the chromosomes of malignant cells and promote chromosomal mutations, just as the loss may do so in aging cells. Thus, the mutations found in malignant cells may have something to do with aging processes but not with cancer initiation. If so, this certainly would not be the first time that most scientists in a field confused effect for cause.

If chromosomal and gene alterations are not the cause of cancer, they do have an important consequence in the development of malignant cells. Cell differentiation cannot proceed normally in offspring malignant cells because the chromosomes received from the parent cell were chemically and structurally altered. This means that malignant cells can never transform back to normal. The secondary chromosomal changes that occur after an initial, epigenetic cancer-causing event are heritable, and the malignant phenotype is permanently fixed in tumors.

Genetic material is not the only possible target for carcinogens and radiation. The chemical carcinogens and forms of radiation that initiate cancer often interact with and damage cellular proteins and RNA. Some carcinogens even react more strongly with proteins and RNA than with DNA. Additionally, many chemicals that cause cancer in rodents are not mutagens, meaning they do not damage DNA. The target of such carcinogens, then, must be epigenetic—outside genetic targets in the chromosomes of cells.

The medical and scientific literature contains many studies providing evidence that the primary defect in cancer is not chromosomal. These observations are usually disregarded because they are not compatible with what most scientists want to believe. For example, in early stages of leukemia, the patient's blood cells can exhibit normal chromosomes. Chromosomal abnormalities come later, in association with clinically evident leukemia.

Marked chromosomal abnormalities can also be found in healthy cells. Many survivors of the atomic bombs of Hiroshima and Nagasaki exhibited pronounced chromosomal mutations of their bone marrow cells and lymphocytes for more than thirty years after exposure to ionizing radiation. Yet these people remained cancer-free.[4]

On the one hand, human cells function normally with many sorts of chromosomal abnormalities. On the other hand, human cells do not require "oncogene" or "suppressor gene" mutations to become malignant. Therefore, the genetic defects that are present in malignant cells are probably the *result* of cancer rather than the initiating events.

## THE "SINGLE HIT" THEORY

The experimental work of the Russian pathologist Dr. Kazymir Pozharisski was discussed earlier. In his paper on intestinal cancer, Pozharisski mentioned other experiments, in which only a single exposure to a carcinogen produced intestinal cancer in a few rats. Since the carcinogen was retained in animals for only several hours, it appeared likely that a "single hit" by carcinogen molecules to target molecules within a cell was all that was necessary for cancer initiation.

In the complex communication network in a developing cell, some parts of the system may be more essential than others. The cell may be able to compensate for some disturbances so that normal communication among cells is maintained. Other disturbances might knock out entire circuits, so that genes are improperly turned on (or off), development is blocked, and the cell can become malignant. For example, various cellular products, such as hormones, attach to specific protein receptors on the surfaces of other cells. The union triggers a cascade of chemical reactions in the cytoplasm that leads to the activation or silencing of genes in the nucleus. Single hits by carcinogen molecules to a critical number of cellular receptors might knock out the receptor function, preventing the cytoplasmic chemical reactions, the activation or silencing of certain genes, and the completion of differentiation—and initiating cancer.

The "single hit" model of cancer initiation explains why, in many human cancers, precancerous changes are not found in tissue surrounding tumors. A "single hit" model means that cancer does not have to be the result of a multistep process in which various areas of the affected tissue can be at different stages of malignancy.

This understanding of malignancy is a hard slap at cancer doctrine and the molecular biologists who control it. To them, cancer is a process in which increasing numbers of "proto-oncogenes" and "suppressor" genes become mutated in a multistep progression that leads from a normal cell to a very malignant one. As I noted in Chapter Five, this concept of cancer's progression is a cornerstone of modern cancer theory, despite the fact that progression is a feature unique to cell lines, not to human tumors.

Widely differing events can prompt identical changes in cellular development. Abnormal differentiation will occur not only when genes are mutated—with a permanent loss of normal function—but also when genes are not "turned on" at the appropriate time—with no permanent mutation. The consequences can be the same in both situations—a breakdown of orderly genetic activity within a developing cell, and possibly the initiation of cancer.

If the second scenario (genes not being "turned on") is the predominant setting for cancer initiation, the epigenetic theory of cancer is more correct. If the first scenario (genes being mutated) predominates, the somatic mutation theory is more correct. However, to ignore the epigenetic theory simply because it does not coincide with the ideas of most scientists is the height of scientific irresponsibility, particularly when we are dealing with a disease that will be, within a few years, the nation's number-one health problem. Yet this is how the cancer establishment operates in order to maintain the supremacy of its ideas.

Thirty years ago, before the meteoric rise of the gene-centered science of molecular biology, the epigenetic theory of cancer did receive serious attention. In the early 1960s, some experts argued that malignancy was not the result of gene mutations.[5] These scientists cited a considerable body of experimental work on chemical carcinogenesis in animals that indicated a causal association between carcinogens binding to cytoplasmic proteins and cancer initiation. However, this association was eventually dismissed by most researchers because they could not see how alterations of cytoplasmic proteins could be perpetuated in subsequent generations of cancer cells. For a trait to be passed on to cellular offspring, it would have to be encoded in the cell's genetic material in some way. The effect of a carcinogen on a cytoplasmic protein should be limited to the cell in which the reaction occurred. It could not be passed on to the cell's "children."

Today we know that the most fundamental regulators of cellular activity are the signals that a cell receives from the other cells of the body. A seemingly temporary change in the cytoplasm—such as the binding of a carcinogen and a cytoplasmic protein—could therefore have a profound effect not only on that cell's cytoplasm but also on its genetic material. If the cell's "reception" is impaired because of aging or because of the influence of a carcinogen, it could *acquire* gene and chromosomal alterations as a secondary effect. Thus, the genetic changes necessary for the perpetuation of the malignant state may be the result of a cancer-causing event that occurs outside of the nucleus and cuts signal transmission. After thirty years of disfavor, it is time to once again take the epigenetic theory of cancer causation seriously.

To leave this chapter on a positive note, it should be remembered that human cells have an exquisite ability to adapt to insult. They can take a lot of exposure to carcinogens before being pushed over the edge to malignancy. Dr. Emmanuel Farber, the pathologist I spoke with at the 1989 AACR conference (see page 52), believes that many of the cellular conditions that are now considered precancerous may not be precancerous at all, but rather just normal cells doing their best to keep within normal limits in response to the daily pressures of life. Don't we—as large collections of cells—do the same thing?

# 9

# The Politics of Cancer: "The Bottom Line Is Dollars"

*I daily witness the growing frustration of creative investigators for whom research funding has become the central theme of their work.[1]*

Daniel M. Cooper, M.D.,
UCLA School of Medicine, 1989

*The real problem with biomedical research today is lack of accountability.[2]*

Bruce Nussbaum,
Senior Editor, *Business Week*, 1990

*Science is a political institution. And like any political institution, it is driven not by debate, but by power, money and public opinion.[3]*

Alston Chase,
syndicated columnist, 1991

During the early seventeenth century, many of the prevailing beliefs about astronomy were wrong. The complicated theories meant to explain the motion of the planets were a bewildering confusion. Nevertheless, the authorities ignored other ideas because they were afraid of being discredited, of losing face and their privileged positions in society.

Similar situations usually lead to similar results, even though centuries separate them. Today, cancer scientists also scramble for success within a tightly controlled bureaucracy. Scientists are only human, and the desire for security and success is understandable, but the end result of this bureaucratic scrambling is the same as it was 400 years ago—faulty and useless science.

The primary goal of most scientists, just like that of other professionals, is a successful career. In research, success is measured by the number of papers a scientist has published in professional journals. A common joke in the cancer industry is that papers are counted, not read. Job candidates with the biggest piles have the best chance of winning the prize—usually a faculty position at a medical school. Those without a long list of published papers are far less likely to achieve such positions. In cancer research, the rule of thumb is "publish or perish."

This mentality fosters research that is focused more on quantity—of data, of publishable findings, of accepted papers—than on quality. Cell lines are the ideal tool for researchers who are under pressure to achieve results quickly so that they can get into print. A long list of publications equals financial security and status, in the form of government or private research grants and contracts; faculty positions; high-paying industrial jobs; or perhaps a place in the "holy of holies," the National Cancer Institute.

The nature of modern cancer research was made abundantly clear during a conversation I had at the May 1991 meeting of the American Association for Cancer Research (AACR) in Houston, Texas. I discussed cell lines with several researchers, including one from the Fred Hutchinson Cancer Center of the University of Washington School of Medicine. She was surprisingly frank about her views of cell lines:

"They don't make any sense, but they are the favorite model," she said.

When I asked if I could use her name, she emphatically replied, "No!"

"Why?" I asked.

"I have to get my money from the NCI."

"Do you feel that if you speak out, your grants would not be funded?"

"Yes, the bottom line is dollars."

Indeed it is; without money, researchers cannot perform experiments, produce data, or report findings in journal articles. The road to peer approval and career advancement will be blocked. Young researchers work long hours to master the process of applying for research funds. Those who are successful will achieve their first major goal, tenured or permanent positions at prominent institutions. Those who are not often find themselves locked out of not only research but also teaching, because most institutions have a tremendous stake in obtaining research dollars.

When grants are awarded, they are funded at a level of at least 140 percent of what was requested. The additional funds go to the institution where the research will be conducted, ostensibly to cover indirect costs such as utilities and wear and tear on facilities. Financially strapped universities and medical schools have become increasingly dependent on this source of income. Innovative science in the field has largely become a thing of the past. Cancer research is a cash cow.

Throughout history, great scientific advances have come from creative individuals investigating what they thought was important, regardless of prevailing opinion or beliefs. Yet in cancer research today, creative scientists must lie in order to carry out their important investigations. Research proposals that do not conform to the status quo are deemed invalid and left unfunded. Papers that openly criticize the reigning model rarely see the light of day on journal pages. By controlling the purse strings, the cancer establishment also controls the direction of all cancer research, crowding out innovation and real advancement in favor of the status quo. In essence, the system rewards mediocrity instead of excellence. Or in the words of a medical school professor of oncology I spoke with at the 1990 AACR conference,

> Once you've become established, protecting your turf becomes most important. I think the biggest problem is the way we direct research; everybody gets funneled into one way. The peer review process basically says everybody should be doing the same thing, instead of real thinking and trying to make sense. Success is agreeing with the people who are in a position to give you the stamp of approval.

## THE NATIONAL CANCER INSTITUTE

A $100-billion government-academic-medical-industrial complex has developed around cancer research and treatment, and its heart is the National Cancer Institute.[4] The NCI is the largest of the thirteen National Institutes of Health, all of which are within the Department of Health and Human Services. It supports a worldwide program of basic and clinical cancer research activities. What is important at the NCI is also important everywhere researchers and oncologists battle cancer.

Four times a year, prominent scientists come to the NCI to sit on committees—called study sections—to judge the scientific merit of research grant proposals submitted by other scientists. Competition for grants is intense, and only those judged the most meritorious are funded. This process of awarding grant money is called the *peer review system* of research proposals. The system was designed to ensure fair, unbiased evaluations by reviewers who are not bureaucrats but rather fellow scientists from around the country.

Unfortunately, peer review does not work as intended. The scientists selected to serve on the study sections are themselves regularly funded by the NCI and have every reason to maintain the status quo. At the NCI, new ideas are threats to the funding status of supposedly unbiased peer reviewers. Those who challenge the prevailing views of the NCI receive low-priority scores on their grant applications and their careers are derailed by the study sections. If a researcher is to survive in the field, he or she must not criticize the prevailing views.

On the other hand, research proposals of leading, established scientists are usually funded automatically by the NCI, as a reward for their "contributions." The establishment rewards those who promote it. This is the level of success every researcher dreams of. Researchers who have worked for decades to become successful will do whatever is necessary to keep their ideas about cancer from being discredited.

Members of the controlling groups in a society usually deal with criticism by ignoring it. When that fails, they try something else—as when Galileo was arrested for delivering the message that the earth was not at the center of the solar system. The cancer authorities have ignored the evidence refuting cell lines, and they will hate what is written in these pages because it refutes everything in which they believe.

Several years ago, I asked Dr. William Kern, my former boss at the Hospital of the Good Samaritan, to review a paper that I had written comparing the features of human cancer and cell lines. Kern had observed, studied, and diagnosed cancer for decades, and I had learned a lot from him. I called him back two weeks later. He had known that cell lines were commonly used in cancer research, but knew very little else about them. "What do you think of cell lines?" I asked. "They are useless," he replied. What a waste, I thought, as I hung up the phone. A pile of useless cell line cancer papers has been published that reaches to the moon.

In the course of my career as an experimental scientist, I had written forty papers that were published in peer-reviewed scientific journals—including journals such as *Cancer Research*, which emphasize cell line research. On all these occasions, peer reviewers deemed my work and qualifications more than adequate. Yet my paper on the inappropriateness of cell lines as a model for human cancer has still not been published. No journal will touch an article that is openly critical of prevailing cancer doctrine. The cancer establishment is effectively eliminating dissent even as the problem grows larger and more people die each year. Perhaps this is the reason the cancer establishment fears controversy.

A dissenter is basically a critic, and critics should look at things clearly and fairly, pointing out defects. A good scientist should always be a critic, carefully analyzing and peeling away the pulp of mere belief and conjecture to reveal the seeds of true knowledge. However, most cancer scientists today are members of the yuppie generation, and the establishment is very much like a large corporation. In such an environment, people are not encouraged to consider the meaningfulness of what they or others are doing.

Institutions that have recruited important scientists to their staffs are generally considered to have the best programs of basic and clinical cancer research. Yet these researchers have the most to lose if the ideas of the establishment become discredited, and therefore they really wear the tightest blinders. M.D. Anderson Cancer Center of the University of Texas Medical School in Houston is considered preeminent in the development of state-of-the-art treatments through research. Nevertheless, the people of Texas generally recognize admittance to M.D. Anderson for treatment as a death sentence. In light of this, can the research of nearly 600 of the "best" scientists and physicians really be considered innovative?

Employment in an NCI laboratory is a reliable road to success for cancer researchers. At the NCI, grant proposals are unnecessary. Research funds are automatically provided by the government. Young scientists from around the world spend their time on research assembly lines, in laboratories with names like pathology, molecular biology, biological chemistry, molecular carcinogenesis, experimental carcinogenesis, cellular carcinogenesis and tumor promotion, tumor cell biology, virus biology, molecular virology, and cellular or molecular biology. In these labs, where laboratory chiefs can select from among hundreds of different cell lines, young scientists learn fashionable methods of research and establish their reputations by publishing papers that often set the standard for the field.

After several years of such indoctrination, these men and women return to their home countries, where they establish their own laboratories based on NCI doctrine. Just as a former assistant secretary of defense is often recruited for a top position in the defense industry, scientists who have risen to top levels at the NCI are recruited for the best jobs in academia and the pharmaceutical and biotechnology industries. For example, the current director of the NCI, Dr. Samuel Broder, was appointed to his post in 1988 after spending his entire career at the NCI, making his way up the bureaucratic ladder. Dr. Marc E. Lippman achieved prominence in the cancer field through his work with breast tumor cell lines at the NCI. He left the NCI to become the director of the Lombardi Cancer Center at Georgetown University in Washington, D.C. Lippman brought many people with him from the NCI when he was recruited by Georgetown. It takes a lot of money to support a large research group. Clearly, those who have become well known at the NCI continue to make the rules even after they leave.

## THE NATIONAL CANCER ADVISORY BOARD

The NCI is also affected by the actions and recommendations of the prestigious National Cancer Advisory Board (NCAB). The eighteen board members serve five-year terms. New board members are proposed by the NCI, and the candidates are submitted to the U.S. president for appointment. The NCAB's task is to review all aspects of the cancer problem, to monitor the quality of the grant review process, and to guide and advise both the NCI and—through the President's Cancer Panel—the president.

The NCAB meets several times a year at the NCI, in Bethesda, Maryland, in sessions that are open to the public, where they listen to experts. In addition, the members review material in their areas of expertise. There are also closed meetings in which recommendations are approved and passed along through established channels.

The NCAB's advice about research priorities comes from its scientists. In 1989, there were five, including one retired from the pharmaceutical industry.[5] It is unlikely that the clinicians on the board—professors in departments of surgery and radiology—could evaluate competently the activities of a research laboratory across the street. Even the scientific jargon used in the research laboratory would probably be incomprehensible to them. Clinicians, by virtue of training, experience, and background, are not qualified to appraise research programs. For the clinicians, the methods of molecular biology, such as a Southern blot (or a northern blot or a western blot), could just as well be an ink blot. Others on the board were administrators of one sort or another in the public or private sector. They were certainly not appointed because of their research backgrounds in cancer.

Two of the five scientists on the NCAB in 1989 stood out. One was Howard M. Temin, Ph.D., a molecular biologist and Nobel prize recipient from the University of Wisconsin. The second was Dr. Enrico Mihich, an immunologist, editor of a cancer journal from Roswell Park Memorial Institute in Buffalo, New York, and the president of the AACR in 1988.[6] Both molecular biology and the immunologic approach to cancer rely on cell lines for their existence and success in the laboratory. The belief that altered genes cause cancer became dominant because of research conducted on cell lines. The belief that the immune system fights cancer is popular for the same reason (more about this in Chapter Eleven).

Having prominent establishment scientists review the national program of cancer research—scientists who were proposed for the NCAB by the NCI itself—is like appointing foxes to guard the chicken coop. The monitoring of the National Cancer Program is lax because the NCAB is nothing more than a rubber stamp for the NCI's interests. This means that the buck stops nowhere. Similarly lax monitoring of the savings and loan industry created a financial holocaust.

The NCAB makes its recommendations to the President's Cancer Panel, which reports directly to the president. The panel could have an important role in reviewing cancer policy since it has the ear of the president. All government bureaucracy is bypassed.

The panel has three members. Before his death in December 1990 at the age of ninety-two, the chairman was the businessman Armand Hammer. The second was a retired surgeon. In 1989, his listed address was the Veterans Administration Medical Center in West Los Angeles, although an operator said his name and phone number were not in the hospital's directory. The third member was the president of a private research institute that has received millions of dollars in grants and contracts from the NCI.[7] Such successful establishment figures are unlikely to rock any boats, and certainly not the boat that contains the funding for one's member institution.

## COMPREHENSIVE CANCER CENTERS

Among the more celebrated fruits of the NCI's labors are twenty-seven comprehensive cancer centers located around the United

States. One hundred million tax dollars are spent yearly on cancer center development. Also, a significant portion of the total funding of federal cancer research activities goes to grants and programs initiated by scientists associated with the centers.[8]

The cancer centers reflect the goals, methods, and scientific approaches of the NCI as well as the failure of the establishment approach. Famous scientists, like James Watson—the Nobel prize-winning codiscoverer of DNA's double helix—acknowledge that these units set up to fight the war on cancer have poor reputations.[9] A close look at just one center—the Arizona Cancer Center at the University of Arizona in Tucson—reveals how little these centers or the NCI can ever hope to achieve.

In 1988, total funding for the center was more than $13 million, but only 51 percent came from the federal government. The rest was made up primarily by the state of Arizona, private donations, and endowments and grants from the pharmaceutical and biotechnology industries. Basic and clinical research took up 77.5 percent of total funding, or approximately $10 million.[10]

On page two of the Arizona Cancer Center's Annual Report for 1988 is a picture of its director, Dr. Sydney E. Salmon. Next to the photograph is the text of the Director's Message. The first paragraph reads,

> The Arizona Cancer Center saw advances in the fight against cancer on several fronts in 1988. From new molecular and cellular discoveries at the laboratory bench to improvements in cancer treatment as a result of our clinical trials the range of achievements is exciting.[11]

When I turned the page I found a photograph of a number of petri dishes in which cell lines grow sitting on a shelf of a laboratory incubator. The microscopic films of cells on their bottoms were invisible to the naked eye. Next to the picture in bold, blue letters was the caption "Research findings pave the way for improved clinical treatments."[12]

The symbol selected to represent the achievements of all of the research activities conducted at the Arizona Cancer Center was cell lines. Like their colleagues at the NCI and other centers, the experts at the Arizona Cancer Center get most of their understanding of cancer from cell lines. And like their colleagues, they are extrapolating from those "new molecular and cellular discoveries at the laboratory bench" to human cancer, falsely believing that what works in a petri dish will work in the human body. Can any of us truly believe that these researchers are "paving the way for improved clinical treatments"?

In 1991, Dr. Salmon was appointed to the National Cancer Advisory Board. This appointment is fitting for the director of an institution where the symbol of its progress against cancer is a petri dish.

When thinking of ways to describe the activities of the Arizona Cancer Center, I thought of interviewing scientists who work there. I called the center in the summer of 1989 and spoke to someone who was an acquaintance of someone I knew. We discussed his current research projects. I did not hide my disapproval of cell lines. I called again and asked if an interview might be possible for a book being written about cancer research. He said yes, but asked me to call back in about ten days because he was preparing a talk for an upcoming conference. I waited ten days and called four times over a two-day period. My phone number was left with his secretary, but I never heard from the scientist. I decided then that it would not be productive to interview researchers from the Arizona Cancer Center. They would not cooperate in debunking the cell line model that dominates their work.

In Chapter One, I discussed a major epidemiological study, "Progress Against Cancer?" by John Bailar and Elaine Smith, which gave the lie to the hopeful rhetoric of researchers such as Dr. Salmon and which blamed ineffective research for losing the war on cancer. The study received some publicity, making it difficult for the establishment to ignore. Dr. Arthur I. Holleb was senior vice president for medical affairs of the American Cancer Society at the time. He responded to Bailar and Smith's paper, writing in defense of cancer research:

> Through basic science we are learning about the pro-
> gramming of normal cells and cancer cells; the body's
> immunological interaction with foreign invaders and
> cancer cells; and how protein chemistry is analyzing the
> structure and mechanisms of cells to map what goes on
> at the molecular level in cancer patients.[13]

Holleb used the word *cells* four times in this brief excerpt from his article, which could have been written by Salmon, implying that "basic science" was learning about human cells and human cancers. But as we have seen, all of the much-vaunted basic science to which Holleb referred is in fact about cell lines and has nothing to do with what actually "goes on at the molecular level in cancer patients." Certainly a great deal of information has been gathered, but to what end?

## CANCER AND THE MEDIA

The work of Bailar and Smith, like that of Barrie Cassileth and colleagues, attracted a fair amount of attention in the media, but it could not counter the continuing flood of false hopes emanating from the NCI. For every morsel of truth about the cancer situation, there are scores of extravagant, misleading stories in magazines, in newspapers, and on television. Cancer has become a political as well as a medical issue because the country has made finding a cure a national goal. As a result, the public is being fooled. For example, in spite of Bailar and Smith's extremely gloomy statistics, which showed the cancer-related mortality rate to be increasing, the NCI was promoting its goal of reducing the rate by 50 percent by the end of the century. A major part of this reduction would come from improved treatments.

In 1989, I interviewed Dr. Bailar by phone. He did not seem to mind that he had been labeled a troublemaker and felt that he had done the job he had set out to do. His paper had had an impact and he was pleased. Bailar felt that there were fewer stories in the press predicting an early conquest of cancer and that the NCI was subtly shifting its policies away from treatment to prevention.

All of this may be true, but it is also as open an admission as one is likely to get from the establishment that research has failed to develop effective treatments. The fact remains that subtle policy shifts are not enough to change either the baleful situation at the NCI or the media's coverage of the cancer war (more about this in Chapter Twelve). In fact, the media coverage of the war on cancer remains all too reminiscent of what Americans were told about the military battles in Southeast Asia during the 1960s. The public is being told that victory will come if we only persevere with the bureaucracy's grand design. Despite the true picture of almost abject failure and growing death rates, the war on cancer—as portrayed in the media—is always entering a new and hopeful phase.

For example, in the summer of 1990, an immunotherapy developed by the NCI's most prominent scientist, Dr. Steven Rosenberg, was extravagantly promoted on ABC's 20/20. The story was reported by Timothy Johnson, a doctor who regularly covers medical news on the program. If an informed skeptic trained in research had done the reporting, the story could well have been about dead-end research supported by your tax dollars for twenty years.

During the week of December 15, 1991, to mark the twenty-year anniversary of the congressional act that declared war on cancer, Edie Magnus did a five-part series about cancer for the CBS Evening News called "Stalking the Killer." In the nightly reports, the bureaucracy's grand design for solving the cancer problem was not questioned. On Monday, December 16, Magnus told the country that "there is again excitement in America's labs . . . . Like 20 years ago there are predictions that a breakthrough may happen soon."[14] Apparently the war on cancer was entering its next hopeful phase. However, a skeptic might consider that if all the thousands of scientists were discovering important things about cancer, there would not be more than 500,000 annual casualties after twenty years of research.

The inadequate state of journalistic coverage of cancer research was documented almost twenty years ago by investigative journalist Daniel Greenberg. In 1975, Greenberg wrote an investigative article for the *Columbia Journalism Review* entitled "A Critical Look at Cancer Coverage."[15] At that time, the national war on cancer was in its third year, during which roughly $600 million was spent under the National Cancer Act of 1971.[16] The article was a journalistic exploration of the cancer establishment in which Greenberg found that things inside were not what they appeared to be from the outside. One physician, who occupied a top administrative post at one of the most eminent cancer institutions, told Greenberg that

> The problem is the closed mind of medicine. Orthodoxy prevails everywhere and it is hard to get them to listen to a new idea. I'm convinced that for some cancers, the survival rates were better a decade ago, but don't tell anyone I said that. The official line is that we're making a lot of progress.[17]

The dangers inherent in conveying such a false picture of progress were made clear by another researcher, who told Greenberg,

> There's a good deal of harm because as long as the establishment is persuading the public that results are

being achieved, there isn't going to be any pressure for supporting alternatives to these dead alley lines of research that dominate the program. It's like Vietnam. Only when the public realized that things were going badly, did pressure build to get out.[18]

The scientists who spoke so frankly to Daniel Greenberg in 1975 did so on condition of anonymity. Then, as now, dissent was not accepted in the cancer research community, and retribution—in the form of rejected papers and grants, canceled contracts, and lost tenure—could be swift and sure. Research results that did not support the prevailing theory were minimized or manipulated to do so. One could not and cannot become prominent by challenging the status quo.

This is the world to which cancer researchers must adapt in order to succeed. It is a world in which dubious treatments and technologies are presented as great new hopes, even as the treatments fail and the death toll rises. The treatment hopes of the last several decades, chemotherapy and immunotherapy, have not stemmed the cancer tide and hold no more promise. In the words of one internationally renowned expert on cancer statistics,

Immunotherapy—stimulating the body's own defense mechanisms—is the latest fad, just as chemotherapy was the much publicized hope in the last decade. But immunotherapy has been around the corner for a long time. What it all comes down to is that there is little in the way of new ideas, and when they do appear, the committees that review them are not receptive. They don't want to be contradicted. And the result is that people become reluctant to come forward with new ideas.[19]

Almost a generation later, immunotherapy continues to be the latest fad and chemotherapy the much-publicized hope.

Both still remain failures, as later discussions of these topics will verify. But since the establishment does not want to be contradicted, harsher and more expensive versions of chemotherapy and immunotherapy are promoted in an attempt to placate a public that might grow restless, waiting for something to work.

# 10

---

# Chemotherapy:
# The Great Hope

*There is no simple solution to the screening dilemma. The difficulties will continue until a system can be defined that bears a direct and consistent relationship to the response of human cancer.*[1]

Kenneth M. Endicott, M.D.,
Chief, Cancer Chemotherapy National Service Center,
National Cancer Institute, 1957

*The administration of chemotherapy in one's office has been and will continue to be a lucrative component of the private oncologist's practice.*[2]

David E. Young, M.D., 1992

In March 1990, the Arizona Cancer Center of the University of Arizona Health Sciences Center sponsored a conference on chemotherapy. The site of the conference was Tucson, Arizona, a pleasant city in the Sonoran desert of the American Southwest. In early spring the weather is ideal, with warm, cloudless days and cool, brisk nights. These facts were obviously not lost on the 800 oncologists who made their way from across the United States and seventeen foreign countries to register for the meeting. The oncologists were a captive, ready-made audience of potential customers for the conference's sponsors, a number of pharmaceutical and biotechnology companies.

It was at this conference that Dr. Samuel Broder, director of the National Cancer Institute, presented the depressing, albeit accurate, statistics on cancer treatment that I briefly discussed in Chapter One. The statistics revealed that in 1987 there was a 5.4-percent increase in mortality from cancer, compared to 1973. In those fourteen years, tens of thousands of our finest scientists had spent billions of dollars in their laboratories to produce about a million pages of new information that was added to the cancer research literature. But most of these data were misleading, actually less than worthless, because we lost ground during that time. No wonder the objective of reducing the cancer death rate by 50 percent by the year 2000 has disappeared from the NCI's list of goals.

It is clear from these statistics that state-of-the-art chemotherapeutic drugs for cancer are ineffective, at best. Although chemotherapy has had some (but by no means complete) success—most notably against Hodgkin's disease, testicular cancer, and childhood leukemia—these exceptions to the general rule could no longer justify the national program of research because the combined numbers of these three partial successes make up less than 2 percent of all new cancer cases each year. The biggest killer, lung cancer, and other common cancers show virtually no response to chemotherapy. Even the director of the NCI could not hide the fact that the "great hope" for cancer is a failure.

## THE HISTORY OF CHEMOTHERAPY

In cancer treatment, survival is directly related to finding cancer in its early stages, before it has spread. When caught early, most cancers can be eliminated completely with surgery or radiation. Once cancer has metastasized, the standard treatment is the administration of poisonous chemicals that have very limited therapeutic effects, partly because chemotherapeutic agents attack healthy, normal cells as well as cancerous ones. This means that even the uncommon therapeutic successes can be offset by problems. For example, children who have been cured of cancer (five-year survivors) often have drug-induced troubles, which include growth failure; gonadal, heart, and lung damage; and intellectual impairment. Chemotherapy often kills patients more effectively than tumors do.

The use of toxic chemicals to treat cancer escalated shortly after World War II. During the war, research on mustard gas revealed its outstanding ability to kill living cells. It is particularly lethal to organs in which there is frequent cell division, such as the intestinal tract, bone marrow, and lymphatic tissue. The effects of mustard gas on dividing cells suggested to scientists of the 1940s that it might have a therapeutic use in the treatment of cancer.[3] Some popular chemotherapeutic drugs used today are, in fact, close relatives of mustard gas—which is why they are so poisonous. There are more than sixty of these drugs on the market, and they are taken by approximately 500,000 Americans annually.

Of course, before mustard gas derivatives could be given to cancer patients, they had to be screened to determine which were the deadliest against tumors and which produced the least severe side effects on the rest of the body. The first models on which drugs were tested were transplanted lymphomas in mice. A transplantable tumor is one that arises in an animal and is then transplanted or passed from that animal to another of the same type and to another and so on.

These screens for anticancer drugs were in use at the NCI from 1945 until the mid-1980s. From 1975 on transplants called the P388 and L1210 mouse leukemias were used. The drug-screening program at the NCI has tested hundreds of thousands of natural and synthetic chemotherapeutic agents for their ability to kill transplantable mouse leukemias and lymphomas over forty years. Those considered promising have been tried on cancer patients. The screening program has proved to be a failure because "promising" drugs have rarely been effective in humans.

In 1960, two prominent men in the field, Drs. James Holland and Charles Heidelberger of the University of Wisconsin, expressed their opinions of transplantable tumors: "We doubt the wisdom of accepting biochemical information derived from transplanted tumors when the data are not consistently applicable to spontaneous neoplasms in the same species."[4]

Just as cell lines exhibit drastic changes in response to life in culture, transplanted tumors change their character markedly as they are systematically passed from one animal to another. This marked change in character could easily explain why chemotherapeutic drugs killed the P388 and L1210 leukemias in mice, for example, but not human tumors.

In the mid-1980s, the experts at the NCI finally discarded transplanted tumors as the model for testing chemotherapeutic agents. In a move designed to improve the reliability of screening for anticancer agents, the NCI switched to a group of tumor cell lines as its primary drug screen.[5] In essence, the NCI abandoned one poor model for another useless model of cancer. Deviation from *in vivo* cancer is even greater in tumor cell lines because transplantable tumors are never grown in an artificial culture system.

In 1992, Dr. Larry M. Weisenthal, a member of an international advisory committee that reviewed the drug screening program at the NCI, wrote, "Of the leading causes of mortality in this country, cancer has arguably the most inadequate predictive model systems for preclinical drug development."[6] Weisenthal pointed out that the cell line panel used for cancer drug screening is so far removed from human cancer that it is as if no screening is being done at all. Cancer patients are serving as guinea pigs in trials of highly toxic, life-threatening new drugs because the model that is used to test them in the laboratory is unsuitable for the job.

Even scientists involved in the drug-screening program at the NCI are complaining. Several of them presented a paper at the 1989 American Association for Cancer Research (AACR) meeting in San Francisco in which they criticized the use of cell lines for drug screening.[7] The authors questioned their use because tumor cell lines have lost the differentiated characteristics of all human cancers during their life in culture.

A discussion of drug testing in a symposium paper presented at the 1988 national meeting of the AACR by investigators at the Wayne State University School of Medicine in Detroit illustrated the misunderstanding of those involved in this most important area of research.[8] The investigators observed that drug discovery is basically a "numbers game." The easier it is to test a potential anticancer agent, the more agents can be tested and the higher the probability of discovering an effective anticancer drug. Yet hundreds of thousands of agents have already been tested on the basis of this rationale—in convenient transplanted tumors and tumor cell lines—and we have yet to come up with truly effective treatments for millions of cancers.

Bristol-Myers Squibb, the company that makes Bufferin, Excedrin, and Ban deodorant, makes five of the best-selling chemotherapeutic drugs. In 1989, it had a 40-percent share of the chemotherapy market, with annual sales of about $600 million.[9]

One of its popular chemotherapeutic agents is Cytoxan (cyclophosphamide), a derivative of mustard gas. The side effects of Cytoxan therapy are horrific. Patients experience nausea, vomiting, hair loss, blood problems, loss of appetite, and heart and lung damage. As if the side effects were not enough, Cytoxan can itself be carcinogenic. One cannot help wondering if the treatment is worth it.

Another of the company's top-selling drugs, Platinol (cisplastin), is so generally toxic that it damages nerves and kidneys and causes seizures and hearing loss. Both Platinol and Cytoxan were given to the late actress and comedienne Gilda Radner, who suffered from one of the most devastating forms of cancer, epithelial ovarian cancer. By the standards of the cancer establishment, she received the best of care and did everything right, to no avail. Like the vast majority of women diagnosed with ovarian cancer, she succumbed to her disease, and the world lost a great comic talent.

## THE REALITY OF TREATMENT

The standard treatment for ovarian and many other forms of cancer is called combination chemotherapy—the administration of multiple toxic drugs. The two-year survival rate for women with advanced epithelial ovarian cancer is only 20 percent. That is, 80 percent of the women who undergo this treatment are dead within twenty-four months.[10] Gilda Radner lived a little longer.

The current great hope in chemotherapy is a technique called *autologous bone marrow transplantation*. It is meant to counter chemotherapy's devastating effect on bone marrow cells, which are essential to the formation of blood and immune cells. Before a patient receives high doses of chemotherapy, healthy bone marrow is "harvested" from the patient and stored. The patient then undergoes intensive chemotherapy. At the end of the chemotherapy, the stored bone marrow is infused into the patient, where, it is hoped, the healthy marrow will thrive and produce desperately needed blood and immune cells.

Intensive chemotherapy with autologous bone marrow transplantation is an awesome experience. Patients spend weeks in intensive care in order to control infections and to survive the devastating toxic reactions of the drugs. Not all of them do. In some trials, as many as 20 percent of the patients died not from their disease, but from the chemotherapy. Physicians are trained to keep patients alive, and high levels of poisonous drugs that destroy all of a patient's blood-forming elements keep oncologists very busy. This costs plenty.

The substantial sickness and mortality that result from such treatment—and its astronomical six-figure-cost—must be measured against its effect on the survival and possible cure of the patients who endure it. Data presented at the 1992 national conference of oncologists indicate that this horrific therapy—like its predecessors—has little utility in stopping cancer.

When oncologists conduct clinical trials to evaluate new drugs, they look for tumor responses. If tumors become smaller, the drug is called effective and selected for general use. However, there is a marked discrepancy between ostensible tumor response and actual patient survival. In only about 32 percent of the clinical trials that reported significant tumor responses to new drugs was survival also prolonged.[11] The fact that a given chemotherapy initially kills some tumor cells does not necessarily mean it will prolong a patient's life. At medical schools, many oncologists practice "academic medicine," finding ways to combine scores of poor drugs in new ways for clinical trials. The result is countless published papers that contain empty facts, like measures of tumor response.

In December 1989, I attended the Twelfth Annual San Antonio Breast Cancer Symposium. Some forty papers were presented on the effect of chemotherapy on breast cancer. One paper, presented by a group from Johns Hopkins Medical School in Baltimore, illustrated how far physicians are willing to go to try to make chemotherapy work. The title of the paper included the word *rescue* because breast cancer patients have to be brought back from the brink of death by bone marrow transplantation. The description of the procedure and its effects seemed more like torture than anything else. It involved removal of bone marrow before intensive chemotherapy. After forty days in intensive care and sixteen weeks of receiving drugs, there were few treatment benefits for most patients. The breast tumors were back in control within eight months, and within two years half the women were dead.[12]

Oncologists should not be blamed entirely for these drastic and ineffective approaches to the cancer problem. Chemotherapy does kill the cells of the research models used to test drugs. It will also kill patients. But it does not effectively kill human tumors *before* it kills patients.

In the nineteenth century, bloodletting was considered good medicine by physicians the world over. Chemotherapy is the bloodletting of our day, with the added horrors of toxic side effects and the added price tag of billions of wasted research dollars.

Although many oncologists undoubtedly prescribe chemotherapy because it is one of the only tools available, the recognized inefficacy of these toxic agents should make physicians far more circumspect in their use. In May 1992, Dr. David Young wrote a letter to the *Journal of the National Cancer Institute* decrying oncologists' use of chemotherapeutic agents. He noted that some physicians disregard the severe limitations of chemotherapy "in order to maintain a very lucrative chemotherapy practice," adding that

> Patients are routinely continued on ineffective chemotherapy despite obvious progression of disease. Others are given adjuvant chemotherapy for long periods, even extending to 5 years for high-risk resected tumors, despite a lack of supportive studies. Patients are sometimes treated with doses as little as one fourth of those in recommended regimens in order to facilitate sustained compliance. Furthermore, patients are compelled to undergo weekly complete blood cell counts despite little or no myelosuppression. This kind of unscrupulous routine is self-serving, costly and detrimen-

tal to the patient. It is unsettling that this practice can continue unchecked in the face of efforts at cost containment.[13]

It is practices such as these, along with the well-known ineffectiveness and toxicity of chemotherapy, that prompt many cancer patients to seek unconventional treatments for their disease. Up to 50 percent of cancer patients seriously consider or try alternatives and spend more than $10 billion each year on unconventional cancer cures.[14]

At the 1990 national oncology conference in Washington, D.C., I had a conversation with a medical school professor who also serves as the director of the oncology department of a large private southern California hospital. His candidness during this private conversation stood in sharp contrast to the sort of public pronouncements made by so many oncologists. On the topic of chemotherapy, he said,

> The whole basis of chemotherapy that filled an auditorium the other day at this meeting was work done in P388 and L1210 leukemia [see p. 107]. There is a whole economy based on that now, the chemotherapy industry. And to look at it now, how in the world could we possibly have done that?

"When are oncologists going to get fed up with what the NCI is doing?" I asked.

"Some are," he replied, "but no one wants to say anything too loudly. The medical oncologist who knows that the NCI is not giving us [effective] drugs, well, he can't say that very loudly or he is out of business."

Perhaps everything, even the care of cancer patients, boils down to economics. Oncologists are not given effective tools to fight cancer by those in research, so they use ineffective ones, such as derivatives of mustard gas. Oncologists know that chemotherapy does very little good for the vast majority of their patients, yet they prescribe it. By constantly touting the promise of the drugs that former Vice President Hubert Humphrey called "bottled death" before he died from bladder cancer, oncologists ward off challenges to their expertise. Cancer treatment based on variations of bottled death is big business, and all big businesses have an inherent obligation to guard their business. The public is being taken not to the cleaners, as in the savings and loan fiasco, but to the mortuary.

The story of chemotherapy for cancer is an ongoing tragedy. In the ongoing search for agents that will effectively kill tumor cell lines, researchers around the world may be missing agents that could kill human cancers. Even more horrifying, however, is the very real possibility that one or more of the hundreds of thousands of agents that have already been tested improperly—and thrown away—could have been effective against real tumors in the body.

# 11

---

# Immunotherapy:
# The Great Fad

*. . . the promise of clinical immunotherapy has been encouraged by pressures on scientists to give a clinical orientation to their projects . . . .* [1]

Harold B. Hewitt, M.D.,
King's College Hospital Medical School,
London, 1979

Nature has endowed human beings with an intricate, sophisticated, and delicately balanced natural defense network known as the immune system. This amazing system protects the body from many threats, including the multitudes of microorganisms, such as viruses and bacteria, that cohabit the world with us.

The immune system is complex and only partially understood. It is made up of a variety of specialized cells that work together to identify, isolate, and destroy foreign invaders—from specks of pollen to the most virulent bacteria. The immune system has direct and indirect links with every organ and system of the body and responds to stimuli from all these systems. If something has an impact on the body, it has an impact on the immune system.

When the immune system identifies a foreign invader, it not only attacks but also learns from the experience, so that it responds faster and more effectively the second time it is exposed to the invader. With every exposure to a new invader, specific immune cells, called lymphocytes, produce specialized proteins known as immunoglobulins (or antibodies). Immunoglobulins are designed to recognize and neutralize invaders by identifying and responding to unique characteristics of the surfaces of invading microbes. These unique surface characteristics are collectively known as antigens. The relationship between antigens and their antibodies is quite specific—once formed, low levels of antibodies remain in circulation in the body but are inactive until they come in contact with their specific antigen "partners." This is why hay fever sufferers continue to have symptoms year after year; hay fever antibodies activate the immune system each autumn when they come in contact with the offending pollen.

Physicians have harnessed the immune system's "troops" in the fight against disease since 1796, when the English country doctor Edward Jenner successfully vaccinated a boy against smallpox. Vaccination takes advantage of the immune system's innate learning ability by "introducing" the system to closely related, weakened, or killed forms of disease agents or to components of the pathogens. Because the pathogen is less virulent in these weakened forms, it cannot overcome the body's defenses. Instead, cells of the immune system quickly form the appropriate antibodies, destroy the pathogen, and are left fully loaded to combat the full-strength disease agent later if necessary. The immune response to vaccination is basically a mild version of the normal disease process.

## CANCER AND IMMUNITY

The idea of battling cancer by freeing the hidden powers of the immune system is not new. Early in this century, physicians attempted to fight cancer by injecting killed bacteria into their patients. In the 1970s, immunologists injected living bacteria of a strain called BCG directly into melanomas, a deadly type of skin cancer. It was hoped that once the immune system was activated to kill bacteria, it would kill tumor cells as well. These approaches to cancer treatment had no effect on the growing tumors, however.

The continued enthusiasm for cancer immunotherapy is based on the belief that the immune system has a built-in mechanism to defend against cancer just as it defends against microbes. The immune surveillance theory of cancer was proposed in the late 1950s and gained widespread popularity in the 1960s. Proponents theorize that the immune system naturally recognizes and restrains the development of tumors. A natural corollary of this theory is the hypothesis that people who develop cancer have, for one reason or another, a deficient antitumor surveillance mechanism. Stimulating or supplementing these innate immune mechanisms is the goal of cancer immunotherapy. The question is, do such mechanisms really exist?

In 1986, Dr. Charles G. Moertel, a senior oncologist at the Mayo Clinic, discussed immunotherapy in an editorial in the *Journal of the American Medical Association:* "This specific treatment approach would not seem to merit further application in the compassionate management of patients with cancer."[2] His rather pessimistic statement reflected the harsh reality of immunotherapy. Far from being the ideal cure for cancer patients, it has proven to be yet another expensive, harsh, painful, and ultimately ineffective cancer treatment.

A generation of research and development has yet to yield an effective immunotherapy for cancer. The cost of immunotherapy, like that of autologous bone marrow transplantation and intensive chemotherapy, reaches six figures. Also like chemotherapy, the latest immunotherapy regimens make patients extremely sick. For this reason, immunotherapy is considered for only a selected handful of the healthiest patients. But unlike chemotherapy, in which the side effects are produced by unnatural, toxic drugs, immunotherapy's side effects are caused by natural components of the immune system that are administered to the limit of the patients' tolerance. The sickness, caused largely by leaking capillaries, is awesome and is characterized by fever, confusion, rigors, and severe anemia requiring intensive supportive care. Not all patients survive the devastating, poisonous side effects.

In spite of the heroic efforts, only a few of these once healthy (but now very sick) patients have tumors that respond by shrinking to some extent in size. Just like chemotherapy, however, these initial tumor responses are generally not durable. The tumors usually start growing again, do not respond to an additional course of immunotherapy, and eventually kill the patients.

Despite its long history of failure, no area of cancer research has received as much public attention and hoopla as immunotherapy. Over the past three decades, magazines, newspapers, radio, and television have heralded the approaching new age of successful cancer treatments utilizing the body's own defense system. Cancer patients are told they can heighten this defense by developing mental pictures of immune cells killing tumor cells. A recent book on cancer treatments advises cancer patients to

> ... think of the cancer as fragile and helpless. Think of the white blood cell characters as strong and noble, fighting for survival. Then envision the drama that will end in the destruction and elimination of cancer through the efforts of your natural defenses.[3]

The public has been told that scientists are learning how to make our immune systems destroy our own tumors. Most of us believe that the immune system plays an important role in fighting cancer. I used to.

## IMMUNOTHERAPY

It is hard to miss when the immune system is called into action. In an infected cut, for example, there can be redness, soreness, and an accumulation of immune cells at the site of infection, sometimes forming pus. The immune system is stimulated into action by the antigens on the surfaces of the bacteria growing in the open wound. Antigens are usually proteins or carbohydrates in the cell membrane that serve as the organism's identifying characteristics. Antigens that elicit an immune response in the human body are said to be immunogenic.

The immune system can distinguish self from nonself because the system inherently recognizes and tolerates the unique set of surface antigens of the body's own cells. Since bacterial antigens are foreign (or nonself), the body's immune cells recognize them, mobilize, and attack. Nonself human cells, such as transplanted organs, can also be destroyed when they, too, are recognized as alien by the recipient's immune system.

Thirty years ago, scientists expected to find that the antigens of tumor cells were different from those of the rest of the body. Cancer cells were abnormal, were they not? Thus, their antigens had to be abnormal as well. Normal, self antigens had surely changed into cancerous, nonself, foreign ones, said the experts. Tumor cells were thought to be almost as foreign as bacteria because of their supposed tumor-specific antigens.

However, in thirty years of intensive searching by thousands of investigators, not one tumor-specific antigen has been isolated from human tumors. Even cancer dogma has been forced to budge a little. The literature has generally abandoned the use of the word *specific* when referring to the antigens of human tumors. Instead, the word *associated* is now used, because the establishment, in this instance, finally accepted the truth: Antigens of human tumors are not specific to tumors; they are always associated with normal tissue as well.

The tumor-specific antigens that were thought to be present

on the surfaces of tumor cells, making them recognizable by every immune system, do not exist. This is a crippling blow to the concept of the immune surveillance of cancer. Furthermore, animals lacking a major organ of the immune system, the thymus gland—which contains the lymphocytes that supposedly look for tumors—show no higher incidence of cancer than normal animals. This is another major blow. Finally, tests of immune function in cancer patients have not revealed any immunological deficits.

Pathologists have known for decades that the immune system is not directly involved in most cancers. Pathologists, by using a microscope, can tell if a body's immune system has responded to a tumor. If large numbers of immune cells were attracted to a malignancy, it would be strong evidence that the immune system had perceived the tumor as foreign and was attacking the malignant cells. But in the vast majority of malignancies, there is no such evidence of immune activity.

Immunotherapy diehards point to the increased incidence of cancer in organ transplant recipients as proof of an immune system–cancer link. Transplant patients receive immunosuppressive drugs to prevent the rejection of organs from others and, as a result, exhibit a high incidence of certain cancers. Obviously, the patients' immune surveillance must be weakened by the drugs.

However, some of the powerful drugs used to suppress immune function are themselves carcinogenic. Thus, the reason for the association of certain cancers with organ transplant recipients need not be a result of faulty immune surveillance. Rather, the drugs being given to suppress immune function also have a dangerous side effect in some patients. They initiate cancer.

Let us look at the situation in another way. The immune system is a marvel. It generally keeps us well protected for life from all of the "bugs" that continually try to gain a foothold inside us. Even when we get infected, our immune systems usually win easily. Would it be reasonable to expect such a wonderfully efficient self-protection mechanism to fail so often against cancer?

So if a built-in anticancer immune surveillance system does not exist, how can immunotherapy ever be effective? Obviously it cannot. The billions of public and private dollars spent over the last thirty years have been wasted.

At the 1988 national meeting of the AACR in New Orleans, I had a brief conversation with one of the most prominent cancer immunologists in the country, Dr. Donald Morton, who was then at the UCLA School of Medicine and the Johnson Comprehensive Cancer Center. Morton moved to UCLA more than twenty years ago from the NCI. There he discovered the convenience of using cell lines as a research tool. He has employed cell lines extensively in research projects on immunotherapy for melanoma.

Melanoma and kidney cancer are usually the only cancer types where immunotherapy is still being considered and where its efficacy must be demonstrated. I asked Morton how his immunotherapy results compared with the results of the standard patient treatment. He then reeled off various statistics about survival times, treatment plans, and so on. After considering the numbers for a moment, I said, "The number of successful treatment outcomes in the two groups are not really statistically different from one another." He agreed, but steadfastly continued to maintain that one day immunotherapy for melanoma would work better. Some 32,000 new cases of melanoma are now diagnosed in the United States every year.[4]

Dr. Morton, as well as all the other cancer immunologists, have spent fortunes over twenty years on projects that have produced nothing of practical value. Today, they offer patients little more hope than they did twenty years ago and about the same survival time after the detection of metastatic disease.

The models used for testing immunotherapy protocols are the same as those for screening chemotherapeutic drugs: tumor cell lines and transplantable tumors. Immunotherapy is tested on tumor lines as they grow in culture (or after the cells have been injected into animals) and on transplanted tumors growing in animals. Immunotherapy usually kills these models. Immunotherapy is very effective against lab-produced representations of cancer. It is human cancer it cannot kill.

Advocates of immunotherapy ignore the fact that tumor cell lines and transplantable tumors change. They exhibit genetic drift, which produces phenotypic instability. The normal self antigens of the original cancer cell can become abnormal, non-self antigens and are therefore "recognizable" as alien to immunotherapy. In the body, however, the immunological characteristics of cancer cells are the same as those of normal cells.

One of the experts who has rejected the false promise of immunotherapy is Dr. Peter Alexander, a prominent immunologist from the Chester Beatty Research Institute in England. He did so in 1977, ten years before Dr. Moertel. In a paper that appeared in the journal *Cancer*, Alexander admitted that the most plausible reason for immunotherapy's failure in clinical situations (when it worked so well in the lab) was that the cell lines and transplanted tumors on which immunotherapy is based "were not realistic."[5]

In 1991, Dr. O.C. Scott, of Saint Thomas Hospital in London, wrote of Dr. Alexander in an issue of *Cancer Research*. Scott, who has been observing the work on tumor immunity for more than forty years, said that Alexander displayed "a level of intellectual courage quite unique in the history of this subject. He actually admitted that he had been wrong."[6]

In 1989, at the Twelfth Annual San Antonio Breast Cancer Symposium, one of the featured speakers was Dr. Dan Longo, a program director at the NCI. Longo gave a lecture on "Biological Response Modifiers in Breast Cancer." *Biological response modifiers* is a synonym for immunotherapy. Changing the name disassociates the subject from immunotherapy's thirty years of failures.

Longo was also a coauthor of a paper that was presented at the 1989 national meeting of the AACR. The paper, like so many to come out of the NCI, demonstrated that immunotherapy killed tumor cell lines in culture or after implantation in the peritoneal cavity of mice.[7] It implied that this experimental killing power also applied to human tumors in the body, even though there was really no justification for such an assumption.

Longo's talk was filled with technical terminology that would be comprehensible to researchers in his laboratory but probably not to a large part of the audience, physicians who treat breast cancer patients. This tactic is fairly common, and it worked. Tumor immunology seemed as "dauntingly complex" as Longo said it was. Longo thus avoided questions since the audience believed that he was just as confused as they were. In fact, his basic message was that it will take the experts another century to decipher cancer's "daunting complexities."

Comments made by Longo during his lecture made it appear that even he wondered about the value of his work. Noting that "immunotherapy has been very disappointing," he said that "if this [immunotherapy] were a drug, we'd throw it away," and that "breast cancer has been very disappointing to our blandishments."

Longo even knocked monoclonal antibodies, which have been promoted for more than a decade as the "magic bullets" that will deliver toxic agents directly to tumor cells, killing only them. Monoclonal antibodies are very specific for single antigens. It was believed that monoclonal antibodies could be made to work against tumor-specific antigens, found only on the surfaces of tumor cells. Such antibodies, when coupled with drugs or other poisons and injected into patients, would act as well-aimed bullets, finding their way to the specific antigens of tumor cells and then binding there, allowing the drugs to kill the tumor cells. Of course, there are no tumor-specific antigens. So monoclonal antibodies directed against tumor cells will also deliver poisons to normal cells, killing them as well. Moreover, many tumor cells will always escape being killed by adapting and removing antigens from their surfaces. Normal and malignant cells are both capable of this adaptive survival response.

Longo also noted the crucial fact that the human body mounts an immune response to injected monoclonal antibodies because they are usually foreign proteins made in mouse cells called hybridomas. Mouse monoclonal antibodies are foreign antigens when inserted into the human body, and the immune system inactivates and eliminates them rapidly. In essence, the immune system destroys the "bullet" before the "bullet" destroys the target.

I spoke with Dr. Longo after his talk. During our brief conversation, I told him that it was a mistake to believe that human tumors elicited an immune response in the body since this belief came from experiments with models that were unrealistic. He responded by telling me that he did not think the transplantable tumors Dr. Steven Rosenberg used were inappropriate. I had read the paper in which the immunological characteristics of the transplanted mouse sarcomas were described.[8] Rosenberg's transplanted tumor laboratory model elicits an immune response in mice and therefore is just as inappropriate as the cell lines used by Longo.

One of the most prominent champions of cancer immunotherapy is Steven Rosenberg, M.D., Ph.D., chief of surgery at the NCI. Dr. Rosenberg has believed for more than twenty years that the immune system can be harnessed to vanquish cancer. Stories about his latest angle on immunotherapy continually appear in my morning newspaper. During the course of the first six months of 1989, *Science* devoted a total of five pages to his research accomplishments. In more than twenty years of reading *Science*, I cannot remember another scientist, from any field, who has received this kind of press. Rosenberg's research group also contributed nine papers to the 1989 national meeting of the AACR, a number almost unmatched in the field. Clearly, this surgeon-scientist is preeminent in the field.

Rosenberg and his colleagues carried out the first authorized gene transfer experiment in a human in 1989. A patient with advanced melanoma received cells containing a gene from a foreign organism. The cells were lymphocytes, the patient's own, removed weeks before and cultured with a growth factor called interleukin 2, or IL-2. In culture, the lymphocytes grew and multiplied and were transformed into what Rosenberg has dubbed "killer cells." Into these "killer" lymphocytes the foreign gene was inserted, and then the cells were infused back into the patient. As reported in *Science*, the researchers planned to use the foreign gene as a marker to follow the lymphocytes as they traveled throughout the patient's body looking for tumor cells to kill.[9]

Unfortunately for the patient, this scenario is impossible. Without tumor-specific antigens, our immune system cannot even see a malignant cell, much less attack it. The first authorized gene transfer experiment was useful for public relations but not for treating cancer. It is about time that the immune surveillance theory of cancer was discarded by everyone, including Steven Rosenberg.

Rosenberg's newest angle is called adoptive immunotherapy because lymphocytes supposedly adopt a new characteristic and become tumor killers after being removed from a patient and cultured with IL-2. When reinfused into the patient, the altered lymphocytes will supposedly home in on the patient's cancer cells. At least, this has been Rosenberg's hope.

Also known as cell transfer therapy, this newest version of immunotherapy is little more effective than all the other attempts over the last thirty years. Only about 20 percent of the healthiest patients exhibited responses to treatment, which lasted about eight months. Clinical trials conducted outside the NCI showed regression rates for renal cancer and melanoma of a trivial 13 percent and 12 percent, respectively.[10] More will be said about the disappointments of IL-2 in the next chapter.

The statistics presented by the director of the NCI in Tucson revealed that between 1973 and 1987, the increases in the death rates for melanoma and kidney cancer were significantly greater than the 5.3-percent increase for all malignancies combined. Yet melanoma and renal cancer patients were given immunotherapy during that fourteen-year period and more than a billion tax dollars was spent on making that immunotherapy efficacious. The human mortality data reveal the truth about immunotherapy: It does not work. If the research had improved immunotherapy's effectiveness, melanoma and kidney cancer death rates would not be going up at alarming rates. The continuing commitment to an ineffective weapon in the war on cancer has precluded work in more profitable directions.

It has been difficult to reconcile the obvious fruitlessness and poor science of immunotherapy with the rare clinical studies in which patients seemed to have benefited. One such trial was reported in 1985 in *Cancer*.[11] Colorectal cancer patients were given a tumor cell vaccine. The vaccine, patterned after those used against diseases such as polio, measles, and smallpox, was designed to activate host immune defenses against foreign antigens. Patients receiving the vaccine had fewer recurrences and deaths than patients not receiving it.

Since all the evidence indicates that tumors do not express foreign antigens, tumor vaccines should be useless. Yet here was a paper published in a respected, peer-reviewed medical journal that showed a clear benefit and led to a large nationwide study funded by the NCI. The apparent contradiction was recently explained.

The nineteenth report by the Committee on Government Operations of the House of Representatives dealing with scientific misconduct and conflict of interest in biomedical research was issued in September 1990. An investigation of the 1985 *Cancer* paper had been conducted because several people associated with the study believed that data were falsified by selective reporting and manipulation. Whether the problem was one of a poorly conducted, biased clinical trial or one of conscious fraud was not decided by the committee. But it was absolutely clear in the report that the results of the study were unreliable.[12]

A consultant who reviewed the matter and found the study seriously flawed also indicted the entire medical profession. He said that the tumor vaccine study was "no worse than hundreds of clinical trials that I find reported in the peer reviewed respected medical literature each year."[13]

The strategy for developing chemotherapy and immunotherapy for cancer is to discover differences between normal and malignant cells and turn the difference into an advantage. Find the weak spot and then go in for the kill. The problem with this strategy is that the weak spots of cell lines and transplantable tumors are not the weak spots of human tumors. Hence, the net result of more than a generation of highly publicized, well-funded research is immunotherapy that will cure mice with cell-line and transplanted tumors but not humans with real cancer.

Two of the more than 500,000 Americans who died of cancer in 1991 were the television producer and actor Michael Landon and the actress Lee Remick. Landon died on July 1, at the age of 54. Remick died one day later, at the age of 55.

Landon had been diagnosed three months earlier with cancer of the pancreas.[14] Remick had been diagnosed with kidney cancer in 1989. She received IL-2 immunotherapy at the NCI, which she called "drastic and horrible and successful," according to a newspaper article that appeared at the time of her death.[15] However, before dying, both Remick and Landon abandoned all treatment. Both came to realize that conventional treatments are "drastic and horrible" but far from "successful."

# 12

---

# Unmasking Biotechnology

*Research in molecular genetics has become a vast multibillion dollar commercial undertaking . . . .* [1]

Michael Crichton,
Jurassic Park, 1990

*A string of product disappointments has fostered an attitude of skepticism when it comes to biotech-product projections.* [2]

Denise Gilbert, Ph.D.,
biotechnology analyst at Smith Barney,
December 1992

The United States spends more on health care than any other nation: 12 percent of the gross national product, or $650 billion in 1991. Cancer consumed about 15 percent of these health care costs. By the year 2000, it is likely that the nation's total health care bill will be $1.7 trillion.[3]

Critics of American medicine blame some of these skyrocketing health care costs on the explosive growth of high-tech medical procedures and treatments. In the world of medical technology, "advanced" is always equated with "improved." In the field of oncology, this belief is patently false. State-of-the-art treatment advances almost invariably mean treatments that are more expensive, harsher, and more difficult to endure. Indeed, for cancer patients, "new" very rarely means "improved."

Nonetheless, pharmaceutical and biotechnology companies spend billions of dollars each year on developing and marketing new "treatments" for cancer. And every year, cancer patients and their insurers spend billions of dollars buying these wares, and the various drugs needed to combat their side effects, in desperate but ultimately futile bids simply to stay alive. For every toxic drug, there are at least three other medications that are used to counteract the devastating side effects. Cancer may indeed be the plague of the twentieth century, but it has been a profitable plague indeed.

## THE MAKING OF A BIOTECHNOLOGY COMPANY

Capitalizing on the seemingly limitless promise of gene splicing, also known as genetic engineering, more than a thousand biotechnology companies sprang up like mushrooms during the 1970s and 1980s. Sporting names like Genentech, Cetus, and Oncogen, some of these companies sought to harness recombinant-DNA technology to develop protein-based products that would cure cancer and make their investors rich. Their goal was (and is) to tap into the enormous cancer-treatment market, a market where the price tag on a new bioengineered drug can be $75,000 for a six-month supply for a single patient.[4]

The rapid growth of biotechnology companies in the past two decades has been a testament to the powers of entrepreneurial skill and Madison Avenue and Wall Street hoopla. To understand just how a biotechnology company becomes involved in the cancer industry, consider the case of a fictional corporation I shall call Protek.

First, the company must identify gene-encoded cellular proteins that have commercial potential. For the purpose of this example, let us say that Protek's scientists have identified a protein called tumor-killing factor, or TKF, which they think will destroy cancer cells. The management of Protek is as enthusiastic about TKF as its scientists are. They decide to use gene-splicing technology to manufacture large amounts of TKF, prove its benefit as a cancer treatment, and obtain a license from the Food and Drug Administration to market TKF to cancer patients.

Protek's first step is to clone, or produce many identical copies of, the TKF gene from an experimental system that makes TKF. The gene-cloning process begins with the isolation of messenger RNA from the human cells that are making TKF as they are grown in the experimental system. TKF messenger RNA is a copy of the DNA of the TKF gene and is used by the cells' biochemical machinery to make TKF.

In a test tube, the messenger RNA is used to construct complementary DNA, with the aid of several enzymes. The DNA is then spliced into bacteria, where it is integrated into the bacteria's genetic material. This product of genetic engineering is called *recombinant DNA* because the bacterial DNA carries a human gene. As colonies of bacteria are grown, the TKF gene is also reproduced. Every one of the millions of bacteria carries the gene for producing the tumor-killing factor.

After spending perhaps $150 million over six years, Protek now has a large supply of bacteria that are continuously producing large amounts of human TKF. If TKF works in patients, Protek's substantial investment in research and development will pay off and the company will make fortunes for its investors, while giving society an effective cancer drug. It is an ideal situation—if only it worked.

In the real world, biotechnology companies have failed in their mandate to find effective treatments for cancer. The biotechnology companies have put most of their eggs in the

immunotherapy basket. They have made enormous invest-
ments in interleukins, growth factors, monoclonal antibodies,
and tumor vaccines. Given what we know of the effectiveness
of these methods, it is not surprising that many of these companies are suffering large financial losses.

## BIOTECHNOLOGY AND CANCER

One of the first great hopes of the biotechnology companies was
a protein of the immune system called interferon. A 1981 book,
*Interferon—The New Hope for Cancer*, stated on its jacket that interferon "could be the major breakthrough in the treatment of cancer."[5] Clinical trials with recombinant human interferon, manufactured by Hoffman–La Roche, began in 1981. Eight years later,
the story was very different. In a lead article in *Cancer Research*, a
prominent immunologist wrote that "there is little evidence of
activity in the common cancers" for interferon.[6]

On November 25, 1985, the cover of *Fortune* magazine
screamed of the new "Cancer Breakthrough." Below the headline was a picture of a bottle labeled "Recombinant Interleukin
2," described on the cover as "Cetus Corporation's tumor zap-
ping Interleukin 2." The first paragraph of the story ended with
the sentence, "So powerful are the new weapons that many
clinicians believe the odds in the struggle against cancer will be
tipped in favor of the patient."[7]

The primary promoter of IL-2 is Dr. Steven Rosenberg of the
NCI. His use of IL-2 to convert lymphocytes into tumor cell
"killers" was discussed briefly in the last chapter. Less than a
month after the *Fortune* article, on December 16, Rosenberg
appeared on the cover of *Newsweek*. In the words of the then
director of the NCI, Dr. Vincent De Vita, Jr., IL-2 "represents
the most interesting and exciting biological therapy we have
seen so far."[8] With the public so hungry for anything that
worked, word of the new treatment spread quickly to all three

television networks. Their nightly news programs cited the research.

By September 22, 1986, IL-2's promise for cancer was being hailed on the cover of *Business Week* as "The New War on Cancer." The magazine recommended a number of industrial developers of immunotherapy as attractive players in the new cancer war.[9] Several of the companies were themselves new. Each had raised more than $100 million in capital on promises that biotechnology would cure cancer. According to *Business Week*, companies like Cetus, Centocor, Xoma, Immunex, Genetics Institute, Hybritech, and Oncogen were poised on the brink of great profits. The first real breakthrough in cancer therapeutics in thirty years was supposed to have occurred.

Cetus Corporation was founded in 1971 and focused on the development of pharmaceutical products to treat cancer. In 1986, Cetus was a top-tier biotechnology company, with corporate offices and research facilities on the east side of San Francisco Bay.[10]

In November 1988, Cetus filed a product license application with the FDA to use IL-2 in patients with metastatic kidney cancer. The company limited its claims for IL-2's effectiveness to this one type of cancer. At the end of July 1990, Cetus came before an advisory committee of the FDA to present data backing up its product license application. The committee did not recommend approval. Seven medical doctors and one Ph.D. scientist were not convinced that IL-2 produced any beneficial effects in renal cancer patients.[11]

After the FDA advisory committee meeting, Robert Fildes, the president and chief executive officer of Cetus, resigned. The company said it would cut its 950-person work force by more than 10 percent. Cetus's stock went from $40 in 1986 to $16 before the committee meeting and then slid to less than $9 afterward.[12] According to the September 1990 issue of *Bio/Technology* magazine, "Wall Street confidence in management has been shaken."[13] In fiscal year 1991, Cetus lost $75.2 million and its investment in IL-2 was over $120 million.[14] The rejection of IL-2 by the FDA had so battered Cetus that it could not survive; it finally agreed to be purchased by another biotechnology company, Chiron Corporation. The stockholders of Chiron approved, and Cetus was bought for approximately $600 million in stock at the end of 1991.[15] Cetus now exists only as a division of Chiron. Other biotechnology players in the new cancer war also continue to lose money. Most have never had a year in the black.[16]

In 1991, Merck Corporation, the world's largest producer of prescription drugs, was rated by *Fortune* as America's most admired corporation for the fifth year in a row.[17] It and other major pharmaceutical companies, such as Hoffman-La Roche, Schering-Plough, Eli Lilly, Bristol-Myers Squibb, and Lederle, have also invested significant resources in trying to make cancer immunotherapy work. Yet their efforts, just like those of Cetus, have failed.[18]

After six years of hype, the headlines about IL-2 have stopped appearing in magazines like *Fortune, Newsweek,* and *Business Week.* The promise of IL-2 had come to an end, the same fate experienced by interferon earlier.

How did the cancer establishment react to the bursting of yet another treatment bubble? At the 1991 annual AACO meeting in Houston, Dr. Steven Rosenberg received the Twenty-Second Annual David A. Karnofsky Memorial Award. Rosenberg was commended by oncologists for his important contributions to immunotherapy. This despite the fact that his immunotherapy approach, IL-2, had, less than a year before, been denied approval by the FDA.

More than fifty journalists attended Rosenberg's press conference on the day he received the Karnofsky Award. Their stories, which appeared in newspapers and on television and radio stations all over the nation the next day, were reminiscent of those that made headlines six years earlier. Clearly, although the promise of IL-2 has come to an end, the hype about immunotherapy has not; most health reporters do not have extensive scientific backgrounds and tend to take the words of scientists at face value. Woodward and Bernstein they are not.

At the press conference, which I attended, Rosenberg spoke of immunotherapy's future:

> This is not like putting a man on the moon. The major obstacle is the immense complexity of the malignant process . . . . We do not understand things we need to understand.

After trying for more than twenty years to reach the "moon" of effective immunotherapy, Rosenberg still does not understand how to get off the ground. There is still no proof that immunotherapy leads to long-term control of cancer in even a few patients.[19] Nevertheless, he remains committed to the idea that the immune system can be made to play an important role in controlling malignancy. Such steadfast faith may be admirable in a medicine man, but not in a scientist. As Francis Crick—codiscoverer of the double helix structure of DNA and a recognized genius of modern science—says in his personal treatise on scientific discovery, *What Mad Pursuit,*

> It is amateurs who have one big bright beautiful idea that they can never abandon. Professionals know that they have to produce theory after theory before they are likely to hit the jackpot. The very process of abandoning one theory for another gives them a degree of critical detachment that is almost essential if they are to succeed.[20]

## BIOTECH'S FATAL FLAW

Interferon and IL-2 are members of a class of proteins called lymphokines. They are produced by cells of the immune system and regulate immune cell function. Interferon has antiviral activity. Interleukin 2 stimulates the proliferation and activity of lymphocytes. If the immune system can be marshaled to fight cancer, the experts know that its lymphokines, such as interferon and IL-2, should help because they are substances produced by activated immune cells that stimulate others into action. However, since all cancer immunotherapies, including interferon and IL-2, have failed, it is reasonable to conclude that the immune system cannot be "turned on" to attack tumor cells because cancer cells are misbehaving body cells, with only normal, nonforeign self antigens.

The pharmaceutical and biotechnology industries have taken the same misstep in drug development as the NCI. They have assumed that the results of cell line and transplantable tumor studies are applicable to human cancer. Industry, like academia, has placed too much emphasis on employing technology and too little on understanding the pathology of human cancer. The managers of pharmaceutical and biotechnology companies, like their scientists, automatically adopt the grand designs of Dr. Rosenberg and other experts of the establishment. Thus, they employ technology inappropriately and produce useless treatments.

Given this state of affairs, it is not surprising that biotechnology's pursuit of effective and profitable anticancer drugs has failed in the marketplace. Unsound science will never lead to effective cancer drugs. Biotechnology analyst Robert Teitelman has described the financial aspects of biotechnology as a "world of fashion and hype."[21] The same can be said for some of the science of biotechnology.

Physical scientists spent $1.6 billion over a dozen years to make a space telescope that does not work properly, and the mistakes made headlines. The National Aeronautics and Space Administration (NASA), a gigantic federal bureaucracy with a lot of visibility, got plenty of heat. Since the end of 1971, the federal investment in cancer research has been well over $20 billion, and the country has a rising cancer death rate. The mistakes here do not make headlines because of a passive press and the fact that biological scientists work behind laboratory doors, using a technology and language that are foreign to most people, including journalists. Is it not time to turn up the heat here as well? Which is more important: Seeing the cosmological Big Bang at the beginning of time or having more time after the biological Big Bang hits one of your cells?

Although Protek's tumor-killing factor is imaginary, tumor necrosis factor is real. It, like interferon and IL-2 before it, is a protein of the immune system that is being hyped by the cancer establishment and the biotechnology industry as a promising anticancer drug. Tumor necrosis factor was first described in 1975 as a serum factor that caused the deaths of transplanted tumors and tumor cell lines in research laboratories.[22] Interferon and IL-2 became the rage in the 1970s and 1980s because they did the same thing. If I were to invest in tumor necrosis factor, I would go short.

Although interferon exhibits little activity against cancer, the FDA approved its use in some cancer patients. In January 1992, the advisory committee of the FDA, which did not recommend approval of IL-2 in July 1990, reversed itself and approved IL-2 for use in patients with advanced kidney cancer. Understandably, dying people will try anything, and they become very effective lobbyists. The lack of an effective drug for renal cancer means that a poor one will reach the marketplace. Only in the pharmaceutical and biotechnology industries does the consumer, in essence, subsidize a research and development pipeline that produces virtually useless products.

# 13

# Every Woman's Nightmare

*There were huge circles under my eyes and I was as thin as anyone you ever saw in documentary footage of Auschwitz. I looked like I was going to die . . . . It was the first time I actually feared what might happen to me—what was happening to me. Who was that person in the mirror? Whose image was reflected? This is a nightmare. Why me? This is cancer! I am only forty years old and I could die.*[1]

Gilda Radner,
*It's Always Something*, 1989

*The twenty year war on breast cancer, which dates from December 1971, has been ultimately unsuccessful.*[2]

Ted Weiss,
U.S. House of Representatives, 1991

*The widespread and unfounded belief that there have been great leaps forward in the treatment of breast cancer is fostered by doctors themselves, by the media and by research organizations and charities which have to exploit the tiniest chinks of hope to keep the money coming in.[3]*

Deborah Hutton,
health journalist, 1993

In May 1989, comedienne Gilda Radner died of ovarian cancer. She had waged a two-and-a-half-year battle against the disease, availing herself of both the most advanced standard treatments and many alternative techniques. Despite her valiant struggle, Radner—like 12,000 other American women that year—finally succumbed to the cancer that began in her reproductive organs.

That same year, 44,000 women died of breast cancer. It has been estimated that one in eight American women will get the disease sometime during her life and that a woman dies of breast cancer every twelve minutes.[4] Tens of thousands more undergo physically and emotionally devastating surgery.

Breast and ovarian cancers are probably the most terrifying diagnoses for any woman. When Radner was told of her malignancy, she reported that "a flush went through my body and out of my mouth came a sound like a guttural animal cry . . . like someone stabbing a knife into me."[5] Most women know the statistics of breast and ovarian cancers—the dismal survival rates, the awful treatment protocols, the painful and disfiguring surgeries. For many, a diagnosis of breast or ovarian cancer is tantamount to a death sentence.


## BREAST CANCER

For thousands of years, the only cure medicine has offered for breast cancer is complete removal of the tumor by surgery before the spread of tumor cells has occurred. Women with metastatic disease are incurable, and state-of-the-art chemotherapy does not improve their median survival of about two years.[6] Even the experts admit they are no closer to a cure for breast cancer today than they were decades ago. Is there any wonder why women today have gotten angry?

A growing number of women have become activists, demonstrating at breast cancer conferences and lobbying Congress for increased funding for breast cancer research. In the December 10, 1990, issue of *Newsweek*, a story described this increased militancy on the part of women who are demanding that something be done.[7] However, it is not the amount of money but how the money is spent that will make the difference in developing cures for cancer. Breast cancer demonstrators should be carrying signs at the NCI reading, "WE DEMAND RELEVANT RESEARCH!" and "THROW AWAY BAD MODELS!"

Although breast cancer research has been an active area of inquiry for a long time, little useful knowledge has come from it. Funding by the National Cancer Institute for breast cancer increased from $33.9 million in fiscal year 1981 to $133 million in fiscal year 1992. This compares with NCI's 1992 estimate of $28 million for prostate cancer and $83 million for lung cancer.[8]

Thousands of basic and clinical scientists spend their entire careers studying breast cancer. A journal, *Breast Cancer Research and Treatment*, covers all aspects of the disease. Several annual conferences that attract researchers and oncologists from all over the world are devoted entirely to this one subject.

One such meeting is the San Antonio Breast Cancer Symposium, sponsored by the Cancer Therapy and Research Foundation of South Texas, the American Cancer Society, and the University of Texas Health Science Center at San Antonio. I decided to attend the twelfth annual symposium in December 1989, even though the paper that I had submitted for presentation was rejected by the meeting's organizers. Since San Antonio's medical school is very active in breast cancer research that depends on breast tumor cell lines, I was not surprised by the paper's rejection. In the paper, I had discussed all of the reasons why the characteristics of the MCF-7 line should not be assigned to breast cancer. Derived from a breast tumor at the Michigan Cancer Foundation in Detroit in 1973, MCF-7 cells have contributed greatly to breast cancer dogma.[9] All of the sessions at national and international conferences concerning breast cancer are dominated by work on MCF-7. The San Antonio Breast Cancer Symposium was no exception.

About a month before the meeting, I received the symposium's program, which listed 219 papers to be presented. About one-half of the research projects were inspired by information that came directly from cell lines. Approximately forty papers discussed chemotherapeutic agents—drugs that owe their clinical interest to their efficiency at killing the tumor cell lines and transplantable tumors that are used to identify "promising" anticancer drugs. Oncogenes were to be discussed in fifteen papers. Their popularity rests on their effect on supposed normal cell lines. Many of the approximately thirty papers on hormones and breast cancer were inspired by the thousands of experiments in which hormones have been added to cultures of breast tumor cell lines. Most of the thirty-odd papers about the biology of breast tumor cells were really about the biology of unstable and undeveloped breast tumor cell lines.

There was a positive side to the meeting. The remaining half of the 219 papers were to deal directly with breast cancer in women. Pathologists, surgeons, and oncologists were to discuss the latest advances in diagnosis, surgery, and hormonal treatment of breast cancer (see Hormones and Cancer, page 143).

## THE SAN ANTONIO CONFERENCE

The conference in San Antonio began with a cocktail reception sponsored by Adria Laboratories, the American distributor of a new breast cancer drug called toremifene. Toremifene had been developed by Farmos Group Ltd., in Turku, Finland, a

# Hormones and Cancer

*In cancers of the breast, the ovary, and the lining of the uterus (the endometrium), the rate of incidence slows after menopause. This indicates that these three cancers depend to some extent on hormones, and they account for 40 percent of all newly diagnosed malignancies in women in the United States. Combination oral contraceptives reduce the incidence of ovarian and endometrial cancer because they alter hormone levels in the body in a beneficial way. The changed hormonal milieu, which reduces ovulation and prevents pregnancy, also reduces the proliferation rate of endometrial cells and cells that line the surfaces of the ovaries. With fewer cancer targets (developing endometrial and ovarian epithelial cells), the incidence of endometrial and ovarian cancer is reduced. Combination oral contraceptives, as currently formulated, do not protect against breast cancer.*

*In adolescent girls, the onset of ovarian function is signaled by the onset of menstruation and breast development. When functioning ovaries produce estrogen and progesterone, the hormones are released into the blood; from blood, estrogen and progesterone molecules diffuse into breast tissue, where they attach to specific receptors on the surfaces of breast cells. These hormonal signals eventually turn genes "on" in breast cells, prompting breasts to grow and develop. In adults, normal breast epithelial cells—those susceptible to cancer—continue to depend to some extent on estrogen and progesterone for their survival.*

*Increased breast exposure to estrogen and progesterone during life—either through early onset of menstruation or late menopause—independently increase the risk of breast cancer. This risk is reduced when the onset of menstruation is delayed or when menopause is early. Thus, estrogen and progesterone, natural body substances, can be considered breast cancer risk factors because they stimulate breast cells to grow and divide. As a result, an increased number of cancer targets, or developing cells, are present.*

*The data on hormones and cancer are not limited to females. In a similar way, the male steroid hormone, testosterone, may contribute to cancer in males because it stimulates prostate cells to grow and divide as well as promoting general tissue growth, including an increase of skeletal muscle mass. The anabolic steroids taken by body builders are derivatives of testosterone and are also suspected carcinogens.*

*Humans pay a price for their gender. Certain body cells are exquisitely sensitive to, and continuously stimulated by, sex hormones. As a result, the rates of cancer of the prostate and breast are among the highest for any organ.*

*In 1896, the Scottish surgeon George Beatson showed that breast cancer, just like the normal breast, depends on ovarian function. He removed the ovaries of three women with breast cancer. In all three, the cancer regressed dramatically. Thus, the rationale for the hormonal treatment of breast cancer is based on a sound foundation, results that were obtained in humans.*

relatively new pharmaceutical company. At the time of the symposium, clinical trials of toremifene were underway in Finland. Later that evening, several speakers told an audience of 600 breast cancer experts including, for the first time, approximately 10 from the Soviet Union, about toremifene.

Adria Laboratories wanted toremifene to be approved for use in America. It sponsored the cocktail party to acquaint American doctors with the drug in the hope of paving the way for future sales. Later that evening, as I listened to two Finnish physicians describe the clinical trials, it became apparent to me that toremifene was probably as effective as the antiestrogens currently in use, but no better.

Antiestrogen drugs such as toremifene work by competing with a woman's natural estrogen for estrogen receptors in breast and other tissues. As a result, they inhibit the action of estrogen. Toremifene and its chemical relatives are structurally similar to estrogen and block estrogen receptors on the surface of breast cancer cells. This prevents estrogen molecules from binding to the receptors, effectively starving the malignant cells of the estrogen they need to grow. Antiestrogens have been used in the United States for more than fifteen years to treat breast cancer, and all clinical trials have showed beneficial results. Women live longer without pain when given an antiestrogen, although their long-term survival is not greatly improved.

The real objective of the cancer industry in general and the pharmaceutical industry in particular became clear to me shortly after the San Antonio conference. After the presentations on toremifene, I spoke with Lauri Veikko, project manager for anti-cancer drugs at Farmos Group, Ltd., who told me that the drug was the result of ten years of research and development by the budding Finnish pharmaceutical industry. Since toremifene's effects were so similar to those of existing antiestrogen drugs, I asked Veikko why his company had put such an enormous effort of time and money into developing a product that seemed to have no real advantage over drugs that were already on the market. He said that Farmos Group had patterned itself after successful American pharmaceutical companies.

I did not grasp the implications of Veikko's statement until several months later, when I read *Health and Healing* by Dr. Andrew Weil, who points out that pharmaceutical companies are always interested in medicines that can be patented.[10] The breast cancer problem is so vast that there is almost an infinite market for antiestrogens. Although toremifene is functionally no different from most other antiestrogen drugs—and is in fact a molecular variant of tamoxifen, the most popular antiestrogen on the market—the Finnish drug qualified for its own patent because it is structurally different from tamoxifen.

The development of closely related drugs by competing firms is one of the pharmaceutical industry's secrets of success. By using a standard formula for its "new" antiestrogen,

Farmos produced a patentable drug that would tap into a large and profitable market. Such maneuvers are logical business moves for companies interested in increasing their profit margin and market share, but they do not produce the innovative treatments for cancer patients that society so badly needs. Indeed, as I noted in the previous chapter, when pharmaceutical and biotechnology companies stray from standard formulas and try to be innovative by exploiting the potential of genetic engineering, they produce ineffective cancer drugs.

Dr. V. Craig Jordan, professor of human oncology and pharmacology at the University of Wisconsin School of Medicine in Madison, was one of the experts who spoke about toremifene in San Antonio. He had conducted experimental studies with the drug for Farmos. In his presentation, Jordan spoke about the MCF-7 line and human breast cancer as if they were identical. One day later I had the opportunity to talk with him about his work. When I told him that human breast tumor cells do not exhibit the progression seen in the MCF-7 cell line, he seemed shocked.

"What do you mean, they don't progress?" he asked incredulously.

"I mean that their character doesn't change over time," I said.

"Do you mean to tell me that when a woman has a well-differentiated breast tumor and dies years later, her tumor is still well differentiated?" asked this professor of "human" oncology.

"Yes, Dr. Jordan, that is exactly right," I replied.

Although he was an internationally recognized expert on breast cancer, this medical school professor could not believe that differentiation is a constant aspect of cancer cells in humans. His understanding of cancer, like that of thousands of cancer "experts," is based not on human cancer but on the unstable and undifferentiated version of cancer that grows in petri dishes.

## THE CLINICAL IMPACT OF POOR RESEARCH

One of the general lectures at the symposium was given by Dr. Norman Wolmark, a professor of surgery at the University of Pittsburgh School of Medicine. Wolmark discussed the primary dilemma that modern medicine faces when the triumphs of technology come up against our ignorance of cellular behavior: "Screening with mammography is possible for every woman, but we do not know how to treat the minimal disease that is detected."

Minimal breast disease is a spectrum of noncancerous conditions associated with cells that look atypical under the microscope. Mammography is sensitive enough to detect many of these atypias. The problem, as Wolmark saw it, was that physicians could not reliably predict which atypias might eventually become cancer. A lack of practical knowledge prevents physicians from always giving the best advice. Since most practicing physicians prefer to err on the side of caution (where malpractice suits are less likely), more mastectomies are probably being performed than are actually necessary.

One example of research that muddles the treatment of breast cancer is the concept of multidrug resistance. Several papers at the symposium dealt with this subject. The December 1990 *Newsweek* article, mentioned earlier, described a test being developed that will check breast tumors for multidrug resistance protein, or MDR.[11] The experts believe that malignant cells elude chemotherapy by using the MDR protein to expel toxic drugs. It is hoped that the test will predict the behavior of actual tumors.

Yet when breast tumors have been examined, less than 15 percent have the MDR protein, whereas all of them elude the effects of chemotherapy. Chemotherapy does not cure breast cancer. Clearly, testing for the MDR protein will provide no useful information. Breast cancers elude chemotherapy because the drugs passed poor screening tests that did not reliably predict how they would work in women. Like almost everything else in cancer research today, the concept of multidrug resistance came from cell lines.[12]

Millions of tax dollars have also been spent on a futile search for breast tumor markers that could separate patients with good prognosis from those whose cancers were likely to recur. Such "prognostic factors" were also discussed in several papers in San Antonio. Of course, because of the fashion of the time, candidates for new prognostic factors are "oncogenes." Hundreds of papers have been written about the amplification and expression of "oncogenes" in breast tumors. Yet all the work has produced nothing useful, according to Dr. Gianni Bonadonna, a prominent Italian oncologist, who stated in a 1992 article in *Cancer Research* that "identification of oncogenes as prognostic indicators remains, as yet, of unproven clinical value."[13]

## VALID EVIDENCE

What is known of breast cancer prognosis has been of value for decades and is due to the work of pathologists. Their low-tech methods have shown that the most useful factors for predicting the outcome of breast cancer are the number of positive lymph nodes and the level of tumor differentiation. The sophisticated, cutting-edge, high-tech methods of molecular biology have added nothing more to our practical understanding of breast cancer.

Although molecular biology has yielded nothing of use in cancer prognosis or treatment, other scientific fields besides pathology have provided several significant clues. Epidemiological studies have shown that breast cancer is often associated with particular risk factors, such as older age at first pregnancy and having a mother with the disease. Yet the sad truth is that most women who contract breast cancer have no known risk factors.

Breast cancer rates vary around the world. Population studies have shown that the United States, England, and the Netherlands have among the highest rates, and the lowest are found in Japan, Thailand, and El Salvador. The difference between the countries with the highest and lowest rates is nearly eightfold. When people move from a country with a low rate to one with a high incidence, the breast cancer frequency in the immigrant group rises to that of the new country. This can happen in as little as a decade.[14]

There is circumstantial evidence that dietary fat contributes to this worldwide picture of breast cancer. Countries with high amounts of fat in their diets have a higher incidence of breast cancer than those countries where less fat is eaten. When people move, they usually adopt the diet of the new country and acquire its breast cancer incidence as well. These findings have led many women to call for more direct research on the relationship between diet and disease. American women would like to know if ten years of a low-fat diet would cut the incidence of breast cancer in half.

Although the circumstantial evidence linking the fatty American diet to breast cancer seems compelling, investigators who have attempted to nail down the case against high fat have yet to succeed. Epidemiologists do know that the timing of and exposure to ovarian hormones (estrogen and progesterone) play a big part in a woman's breast cancer risk. During a woman's reproductive years, the levels of these hormones rhythmically rise and fall over each twenty-eight-day cycle. Breast cancer risk appears to depend on the total lifetime exposure to these hormones.

In some underdeveloped countries, girls who are poorly nourished do not menstruate until their late teens. These same countries also have a low incidence of breast cancer.[15] Two hundred years ago, American women also reached menarche (the onset of menstruation) at around age 17.[16] Today, American women have improved nutrition, reach menarche at an average of 12.8 years, and have an increased rate of breast cancer.[17] It is possible, therefore, that the rising breast cancer rate is due, indirectly, to improved nutrition. A high-caloric diet during childhood lowers the age when menstruation begins, which leads to an increased exposure to estrogen and progesterone during life, which leads to an increased breast cancer risk. Fat per se may not increase risk.

Another reason for the association between high-caloric diets rich in fat and breast cancer may be that fatty tissues, such as the breast, can become storage sites for toxic chemicals and carcinogens. For example, in Israel between 1976 and 1986, there was a dramatic drop of more than 30 percent in breast cancer deaths among young women.[18] This drop was not seen in other nations.

For at least ten years before 1978, milk and dairy products in Israel were contaminated with high levels of three highly persistent chlorinated pesticides, including DDT. All three compounds are carcinogenic and are known to accumulate in body and breast fat and to be concentrated in human milk. For a decade before 1978, Israelis were shown to have high levels of the pesticides in their body fat and milk.

In 1978, all three pesticides were banned. The ban resulted in a large drop of pesticide levels in the bodies of Israelis. Many public health experts believe that the high concentrations of the pesticides in the Israeli diet caused a breast cancer epidemic in Israel.[19] The dramatic change in the environment in 1978 (the result of the pesticide ban) is believed to have been responsible for the large drop in breast cancer incidence over the next decade, especially among younger women.

Some chemicals, including those that were eventually banned in Israel, are not easily broken down by the body and accumulate in fatty tissue. When chemicals of this kind are present in the environment, food animals—from cows to ocean fish—consume the chemicals and store them in their fat. Women who eat contaminated animal fat will, in turn, store the chemicals in their own fatty tissues, including those of the breast. There, especially in the teenage years, when breast tissue is developing, carcinogens could transform a normal breast cell into a malignant one. The higher the levels of carcinogenic chemicals stored in the breasts, the more likely that they will cause a malignant cellular transformation. Thus, it may be more than coincidence that the current breast cancer epidemic in the United States has occurred during a time when pesticide use has reached more than a billion pounds a year and chemical contamination is a major public health concern.

## OVARIAN CANCER

Ovarian cancer is the fourth leading cancer killer of women. It strikes all ages but has a peak incidence between age forty and seventy. About 20,000 new cases are diagnosed annually in the United States, and about 12,000 women die of the disease each year.[20] Like breast cancer, the treatment of most ovarian cancers leaves much to be desired.

Sheila Baker and Jane Kohlrosser are sisters, and both are dying of epithelial ovarian cancer, cancer that began in the surface epithelium of the ovary. Epithelial ovarian cancer constitutes about 70 percent of the cases. The sisters appeared on CNN's *Larry King Live* on April 2, 1991, to complain that their health insurance was not delivering on its promises. Blue Cross and Blue Shield of New York State did not want to pay the $100,000-plus tab for intensive chemotherapy and autologous bone marrow transplantation recommended for each woman by the Dana-Farber Cancer Center of Harvard Medical School. For the advanced stage of each cancer, each sister had only a 20-percent change of surviving two years, even with the standard treatment of combination chemotherapy. Just as in breast cancer, state-of-the-art chemotherapy for ovarian cancer is as ineffective as no treatment at all.[21] Who would not be sympathetic to the women's plight?

When one sister related that they were told by the cancer center that intensive chemotherapy and autologous bone marrow transplantation might cure them, the villain was quickly identified as a cold, heartless, and greedy insurance company. In truth, high-dose chemotherapy with ineffective drugs that are poisonous would probably give the sisters more suffering than benefit. The villain is really a cancer establishment that has not been able to develop one effective treatment for ovarian cancer and all the other common malignancies in more than thirty years.

New York State Blue Cross and Blue Shield is a nonprofit organization. Its representative said on the show that money for unproven therapies does not come from Blue Cross/Blue Shield but from its subscribers, in the form of higher premiums. "They are the ones who would be affected," he said.[22] To avoid being discredited, oncologists continue to try to make chemotherapy effective at all costs, and all of us may be paying for it in one way or another.

Taxol is the newest antitumor drug for ovarian cancer. The director of the NCI, Dr. Samuel Broder, believes that taxol is the most important new chemotherapeutic agent to come along in a decade or more. It was recently approved for clinical use. What can cancer patients expect when the most important new drug in more than ten years is widely available? Not much. Several months' relief for a minority of patients with ovarian cancer, breast cancer, leukemia, and melanoma is the benefit taxol is giving in clinical trials. In an article in the *New England Journal of Medicine,* oncologist Cornelius O. Granai, of Brown University School of Medicine, asked his fellow oncologists why the benefits of taxol are being exaggerated to the general public. "Whatever the explanation," he wrote, "the creation of unrealistic expectations in patients and society is unfortunate and sometimes shameful and it suggests a standard of care we cannot deliver."[23]

Most breast cancers have been present for years by the time a lump can be felt or detected by mammography. Yet long before this time, malignant cells can get into the bloodstream and travel to the lungs, liver, or bones, where they grow into secondary tumors that kill. Ovarian cancer is known for its rapid spread. Early detection will not prevent many breast and ovarian cancer deaths because the biological behavior of the tumor has already been established. Effective treatments for systemic disease are desperately needed. Oncologists who suggest protective mastectomies or protective ovariectomies for high-risk women without breast or ovarian cancer are acknowledging that systemic breast and ovarian cancer cannot be cured. Effective drugs are lacking because scientists are very restricted in their vision. How breast and ovarian cancer are defined depends on what courses have been taken, what books have been read, which journals have been studied, and what research training and practice have been followed.

I wish to point out, as often as the reader will allow, that the object of all cancer research should be to learn about human cancer. Women cancer advocates clamor publicly for more money for research. So far, they do not understand the real problem.

# 14

---

# The Real Picture

*Thus, [experimental] models, while extremely helpful in science, can also be a source of blindness.*[1]

Bernard Barber, Ph.D.,
Columbia University, 1961

As I drove out of Texas after the 1989 San Antonio Breast Cancer Symposium, a thought kept recurring in my mind. The old adage, "Seek, and ye shall find," is remarkably appropriate for the state of cancer research. If scientists look hard enough, they will, eventually, find things in the body that they first observed in a petri dish. For scientists, these *in vivo* finds "prove" that life in culture and in the body work the same way. This belief is spectacularly misguided.

There may be a rare human tumor that exhibits nonself antigens and is immunogenic, but this is just the exception that proves the rule. There may be a human tumor that is effectively treated with chemotherapy, but this, too, is rare. There are human tumors in which a "proto-oncogene" is mutated and there are other tumors of the same type in which it is not. Tumors of another type may have a different mutated "proto-oncogene." Thus, what molecular biologist Dr. Peter Duesberg wrote in 1985 remains true today: "There is still no proof that activated proto-oncogenes are sufficient or even necessary to cause cancer."[2]

Cancer knowledge derived from a petri dish is incorrect because it begins with an unreliable model that bears no resemblance to the *in vivo* environment. The human body, with its trillions of cells of 200 different types, is a living cosmos. This cosmos is regulated by complex communication among cells as they age and differentiate over time. Cells that have been removed from that universe and established in culture do not communicate normally. In the absence of such communication, cells become ageless, undeveloped, and unstable. Such cells should not even audition for the part of a real normal or malignant cell.

In the body, both normal and malignant human cells have fixed phenotypes and tend to be genetically stable, despite the mutations that malignant cells and old cells may acquire. Cells established in culture, however, are grossly *un*stable, with genetic and phenotypic characteristics that evolve randomly over time. Cells in culture do not have the exchange of signals that maintains stability and differentiation. Cell lines, the most popular and influential model of both normal and malignant cells, provide as inaccurate a picture of nature as Ptolemy's geocentric model of the solar system.

In the cancer universe, everything revolves around a plastic dish in grossly unstable orbits that change randomly over time. No wonder most cancer researchers are confused by the complexity of an experimental model that became popular (ironically) because it was thought to reduce cancer to the essentials.

Many of the essentials of cancer are contained within each malignant cell of every tumor. Although cell lines allow direct observations of a large number of individual cells, these cells have descended from one tumor cell grown in an artificial environment completely controlled by the experimenter. Under such conditions, it is inevitable that cells will behave in ways entirely different from the way they would in the body. The secrets of cancer cannot be uncovered by probing deeper into petri dishes, because in real life cells interact and develop in complex ways that cannot be duplicated in long-term culture, and these processes have an inevitable, fundamental impact on malignancy. Failure to recognize this fact has tainted all cancer research.

## USELESS RESEARCH

Cancer research is booming because there is clearly a tremendous need for meaningful answers to this major health problem. Each year increasing numbers of papers are submitted to the annual meetings of the American Association for Cancer Research (AACR). At the 1990 conference in Washington, D.C., 2,700 papers were presented. Since virtually all had a number of authors, these papers represented the work of about 10,000 scientists. The previous year, 2,500 papers were presented in San Francisco.

Most of the papers at these meetings are not about human cancer but about cell lines. Because of time and seating limitations, most papers are presented at daily poster sessions, held in large exhibit halls of convention centers. For poster sessions, scientists bring discussions of their research projects in condensed written form with pictures, diagrams, and charts; these are pinned to assigned movable bulletin boards, set up in rows in an exhibit hall.

For several hours each day, an author must be present at the poster to answer questions. This allows cancer scientists to discuss research in an informal setting, where they can be far more candid than when speaking before an audience. Furthermore, the young members of research teams are often the ones assigned to posters, and they are usually more frank and intellectually open than their older colleagues.

At a poster session at the 1988 meeting of the AACR, I talked with a young scientist about cell lines. After a brief discussion of their limitations, I asked why he used them. He replied without hesitation that "they are a lot of fun."

Cell lines are "fun" because they are convenient. You can do almost anything with them. Cell lines are also unlike any other life form on earth. Scientists enjoy describing their wildly unstable characteristics because nothing of the sort has ever been seen in the history of biology. Unfortunately, however, scientists attribute this instability to human cancer cells as well, despite the reality seen and documented in pathology labs every day. As a result, established views about cancer are as bizarre as the cell line research on which they are based.

In the body, defects that produce disease are usually relatively small and stable. For example, in sickle cell anemia, an alteration of a single gene leads to a consistent alteration of the hemoglobin molecules in red blood cells. In some forms of diabetes, pancreatic cells lose the ability to manufacture a single crucial protein, insulin. A heart attack can be caused by the clogging of just one coronary artery, decreasing blood flow to the heart. The area of dead heart muscle that eventually results can be fatal for all of the cells in the body.

In contrast, cancer researchers tell us that malignant cells become wildly different from normal cells and that their characteristics continue to metamorphose over time. If they were correct, pathologists would never be able to reliably predict the future behavior of a cancerous tumor. Well-differentiated tumors would frequently change, becoming poorly differentiated, faster growing, and more aggressive. In reality, well-differentiated tumors stay that way. For example, pathologists know that many men will outlive their well-differentiated prostate cancers without any treatment because the tumors are slow growing and will remain nonaggressive. Therefore, the mutations that malignant human cells seem to accumulate over time—which are thought to increase their aggressiveness—really have little effect on the behavior of malignant cells. Tumor phenotypes remain fixed over time.

If cancer cells really did routinely change their characteristics, we could never hope to develop effective treatments.

Therapies that were effective one day would become ineffective the next. But few researchers consider this logical downside to their pet theory. The more complex and incomprehensible cancer becomes, the more money will be poured into research to unravel the mystery. As Thomas S. Kuhn pointed out in *The Structure of Scientific Revolutions*, truth is not necessarily a criterion of the advancement of science.[3]

## OCCAM'S RAZOR

In 1987, Dr. Garth Nicolson, chairman of the Department of Tumor Biology at M.D. Anderson Cancer Center of the University of Texas, wrote the following in *Cancer Research*: "The viewpoint that no unitary concept can give a satisfactory explanation of the intimate nature of cancer remains quite valid today."[4] Nicolson, like most of his colleagues, believes that cancer is so changeable that it may yield not one but possibly a thousand different explanations. Compare this view with that of Dr. Eugene Day, who wrote in 1961 of the subtle differences that separate normal and malignant cells. The differences between Nicolson and Day are anything but subtle. They are separated by thirty years in which the field of cancer research has been effectively taken over by cell lines.

Physicists, unlike cancer researchers, recognize the essential similarity between different things in nature and have developed the theory of quantum mechanics to explain it. Their ultimate goal is a grand unified theory that will bring together all forms of physics. Physicists have not forgotten the injunction of the fourteenth-century philosopher William of Occam, who stated that simple theories with the smallest number of arbitrary assumptions are to be preferred.

Cancer researchers, on the other hand, never invoke "Occam's razor." Because their "normal" cell model is undifferentiated and does not age, they do not realize that human cancer results from defects of cell maturation. Therefore, the experts view cancer as many different diseases rather than one species with many breeds. In their dependence on a poor model, cancer researchers fail to see the essential similarities of most malignancies and cannot believe that a "unitary concept" (or unified theory) of cancer is possible.

Modern cancer research begins with the wrong experimental model and goes downhill from there. The methods of molecular biology—for all their power—cannot produce accurate information because of the unstable genetic nature of the cell line model. Scientific knowledge should come from careful experimentation and observation of models that accurately mimic the human body and the developmental pathways of human cells, both malignant and normal.

Ironically, the earliest critics of cell lines were the scientists who first studied them in the 1950s. They noted the essential instability of cell lines and cautioned that "it is of course of paramount importance in experimental biology to employ as nearly uniform material as possible so that repeated experiments will give comparative results."[5] With this simple, straightforward admonition about the critical importance of using accurate experimental models, the early cell line researchers effectively warned their successors against their use. Nonetheless, cell lines took off.

## CANCER AND AGING

Most cancer experts find the idea of a link between cellular aging and cancer initiation incongruous because their model for studying carcinogenesis, "normal" cell lines, does not age. But there are several cellular scenarios that could explain the relationship between aging and cancer.

In developmental zones of organs, cells divide as they differentiate. If, for some reason, partially differentiated cells are left behind in developmental zones, they could accumulate defects over time, finally transforming one of them to the malignant state. It also seems reasonable to assume that stem cells—like fully developed cells—age over time. Some of the body's cells age and die in a matter of days, but stem cells seem to age over a lifetime, as they divide regularly to provide replacements for old or defective cells. Some studies suggest that stem cells finally become old and cease dividing after they have completed a certain number of divisions.[6]

Cells are basically self-repairing machines, and stem cells may age because they lose the ability to cope with wear and tear. If their repair machinery becomes impaired as they divide, stem cells would accumulate damaged enzymes and other molecules over time. Some of these damaged molecules would be distributed to their differentiating offspring, where they could interfere with developmental processes, perhaps increasing the risk for transformation to the malignant state. Thus, the aging process may make the differentiating cells that result from divisions of older stem cells more susceptible to cancer. Or the damaged molecules within the differentiated offspring of older stem cells may make them less able to keep in check a malignant cell neighbor that arises among them. Either scenario could explain at least part of the relationship between cancer and aging.

Cell growth and proliferation is regulated by many layers of checks and balances. One such check on the promiscuous divisions of a malignant cell may be signals that are sent by its immediate neighbors. The experimental work of Dr. Kazymir Pozharisski showed that a malignant cell could be formed after only one administration of a carcinogen. After an administration, there was a long time lag before a few animals developed tumors. Thus, normal aging processes may, at some point, allow a malignant cell within an organ to escape the control of normal cells of the same type that surround it.

It is doubtful that the father of evolutionary theory—the great Charles Darwin—would agree with the most important contemporary belief about the behavior of cancer cells. According to Darwinian theory, natural selection (or survival of the fittest) occurs in nature in response to environmental changes. When the environment does not change, neither do the animals that live in it. Sharks, for example, have been living in the relatively change-less oceans for millions of years. As a result, modern sharks are essentially identical to their ancient ancestors.

A classic example of what some consider to be evolution at work is the phenomenon of "industrial melanism" in peppered moths in England. Prior to the industrial revolution, dark-colored moths were a small subgroup within the peppered moth population, most of which were light. As the trees in industrial districts became blackened by soot, the number of light peppered moths declined and dark forms predominated. The darker moths were less conspicuous to predators when sitting on soot-blackened trees, and thus they survived while their lighter cousins were consumed.

This natural experiment demonstrates the effects of predation on the survival of the dark and light forms of the peppered moth in a clean environment and in one that had been changed by pollution. When the environment changed, an uncommon physical characteristic provided a survival advantage for the moths, and as a result, the moths evolved as that characteristic increased in frequency in the moth population.

A centerpiece of tumor cell biology is the concept that malignant cells evolve in the human body. Yet the body is far more similar to the oceans in which sharks developed than to the soot-blackened neighborhoods of England that spawned increasing numbers of dark peppered moths. The internal milieu of the body is kept constant through the minor adjustments each cell makes in response to the feedback from other cells. Under constant conditions—in the ocean or in the body—where sharks or normal and malignant cells are well adapted to their environments, there is no advantage in changing. In fact, just the opposite would occur: When changed forms did arise by chance, because of mutation, they would probably be selected against because of survival disadvantages and would not survive for long. A malignant cell that "works" in the human body has no reason to evolve. Tumor cell evolution is only a characteristic of cells when they live as microscopic scum on the bottom of a plastic dish in conditions never seen in the human body.

This does not mean that tumor cells do not make adjustments to their structure and function. Not all malignant cells in a tumor are identical at any given moment. There is variation within tumors, but it is the same sort of cellular variation that one sees among normal cells existing in the same organ. All cells, like the humans of which they are very small parts, are capable of adapting to changes in their environments by changing themselves to a limited degree.

The changes seen in normal and malignant cells are physiologic, meaning that they are completely reversible and do not involve mutations of chromosomes or genes. For example, a particular skin cell called a melanocyte adapts to sunlight by producing melanin, a pigment that darkens the skin. During winter, the cells slow down the production of melanin and the skin returns to its original shade. Although the physiologic change may be dramatic, the genetic material of the cell does not change. In contrast, the instability and evolution of cell lines often reflect cellular variation of a much larger magnitude, caused by mutations, changes in the number and structure of chromosomes and genes, that are not reversible.

In biology, form is closely related to function. For example, the structure of a bird's wing is designed for flight. The structure of a DNA molecule determines its mode of replication inside a dividing cell. This means that the large body of structural information in the tumor pathology literature can provide insights into the abnormal behavior of malignant cells. Yet few cancer researchers seek out this knowledge. In their experience, they cannot imagine that human tumors exhibit the limited variability that is characteristic of the stable phenotypes of normal cells.

In ancient Greece, scientists were patrician thinkers. They presided at academies and seldom dirtied their hands by empirically testing the ideas they debated each day. Most of today's cancer researchers also exist in an academic environment, and although they actually test their theories, they use a model that is so far divorced from reality that it ensures their research will remain "pure" and unrealistic. We do not refer to material or beliefs that have no practical value as "academic" for no reason. The schism between academia and the real world has been recognized for centuries.

Although it can take hundreds of years, poor models of nature are eventually discarded when the weight of the evidence becomes impossible for the scientific establishment to ignore. Incorrect theories finally die from their own ponderousness, but in the case of cancer, millions of patients will die before the dogma does.

# 15

---

# The Promise of
Correct Models

*Nobody likes to ask if a model is really correct, since if
they did, most work would have come to a halt.*[1]

Francis Crick, Ph.D.
*What Mad Pursuit*, 1988

In May 1991, a report issued by the American Hospital Association confirmed the severity of the cancer problem. On May 13, the
*Tucson Citizen* trumpeted the report's findings with a bold headline: "By 2000, Cancer Will Be the Leading Cause of Death."

I read this headline as my wife and I were driving to
Houston, Texas, for the annual meetings of the American Association for Cancer Research (AACR) and the American Association of Clinical Oncology (AACO). Not surprisingly, no one
at the meetings mentioned the report, which concluded that
cancer will eclipse heart disease as the primary cause of death
in America by the end of the century.

There was a critical, although implicit, lesson to be learned
from this rather depressing report—one that few cancer researchers seemed to perceive. The battle against heart disease
is obviously being won, even though billions more dollars are

being spent on cancer. The death rates for the two diseases are headed in opposite directions. The drugs and procedures that have been developed to fight heart disease—including those that lower cholesterol and blood pressure, control arrhythmias, and improve cardiac function—actually work in humans. The investment in heart disease is paying big dividends, whereas the larger investment in cancer is not. What are the cardiac researchers doing right?

I was confident I knew the answer. My suspicion was confirmed when I visited a university's science library several weeks later. There I discovered that not one of the thirty-six research projects that were reported in the May 1991 issue of *Circulation* had employed a cell line as an experimental model. The data on the heart and circulatory system had been gathered from animals whose biology mimics that of humans. Rats, mice, hamsters, pigs, and rabbits were sacrificed in experiments, not cells that have lived in petri dishes for years.

The June 15, 1991, issue of *Cancer Research*, in contrast, also contained thirty-six papers—twenty of which were investigations of cell lines. *Cancer Research* publishes about 8,000 pages per year. Most of them have significance only in the ivory towers of academia. All other experimental cancer journals are similar.

The reason for the failure of cancer research is clear, although the experts will deny it. On the one hand, drugs and procedures for heart disease are developed in mammals and, as a result, are also effective in the human mammal. Cancer drugs, on the other hand, result from studies of cell lines and are not effective in humans.


## BETTER MODELS

Thirty years ago, Dr. Eugene Day said that by studying natural tumors in animals, the secrets of human cancer would be revealed, because animal cancer is a good representation of human cancer. Inbred strains of rodents have been developed in which cancer occurs often. Such *in vivo* cancer models might serve as reliable screens in which to test anticancer treatments.

A small biotechnology company in San Diego, AntiCancer Inc., recognizes the need for clinically relevant cancer models in which to test cancer treatments, and it has taken a promising new approach to model development—nude mice. Nude mice are a mutant variety that acquired their name because they are hairless. They also lack a major organ of the immune system, the thymus gland, which means that tissue from other species—animals or humans—will not be rejected when transplanted into nude mice.

AntiCancer Inc. implants small pieces of living, intact human tumors from surgical specimens into the corresponding organs of nude mice. A recent paper described the results of research in which human colon tumors were implanted on the colons of mice.[2] In the nude mice, the colon tumors appeared to retain the original properties of the human tumors. Their microscopic structures and differentiation did not change, and like colon tumors in the human body, they grew locally and some metastasized to mouse lymph nodes and liver. The scientists at AntiCancer Inc. believe that a model of human cancer in nude mice should be superior to existing models, "which use highly deviated tumors to study new treatment modalities."[3]

These scientists make use of the existing supply of human tumor material. In surgical pathology laboratories across the nation, excess portions of many thousands of human tumors, every year, are simply placed in formalin-filled containers and stored on shelves for a time before being thrown away. Who knows what might happen if thousands of skilled scientists discarded their incorrect models and concepts and brought themselves and their powerful methods into such an environment.

## HORMONES

I have often reflected on a few animal and human studies, lost among the hundreds of thousands of cell line modeled projects, that might serve as bases for successful strategies against cancer.

In 1970, an epidemiological study revealed that women who had a full-term pregnancy before the age of twenty acquired a certain degree of protection for life from breast cancer.[4] It has been estimated that if all women bore children before the age of twenty, the incidence of breast cancer would be reduced by as much as 30 percent. Although this is unlikely to occur, an understanding of how an early pregnancy protects might enable scientists to devise a way to protect women at high risk, without pregnancy being necessary.

A group at the Michigan Cancer Foundation in Detroit has been studying experimental breast cancer for a number of years. Their results, obtained with rats, explained why an early full-term pregnancy was protective against breast cancer.[5] In rats, as in humans, an early pregnancy protects against mammary cancer. After an early pregnancy, the mammary epithelial cells susceptible to cancer are maintained at a higher than usual level of partial development for the rest of the animals' lives. We know this confers protection because as cells differentiate, they divide less frequently. Fully differentiated mammary cells are present at the end of pregnancy and during lactation. They are not susceptible to cancer because they have lost the ability to divide.

This very interesting investigation suggests that women's breasts would also be protected if, before the age of twenty, their epithelial cells became fully developed for a short time. After such an event, the cells should be maintained at a higher than usual level of partial development, just as in the rat, and this would protect against breast cancer later in life.

Birth control pills have been around since the 1960s. They contain the right combination and concentration of hormones necessary to prevent pregnancy. A breast cancer reduction pill, containing the right combination and concentration of hormones necessary to bring breast epithelial cells in high-risk young women to a lactational state, simulating for the breasts the result of a full-term pregnancy before the age of twenty, does not seem an unreasonable goal. Moreover, since multiple deliveries also protect against breast cancer, women might benefit even more if their breast cells became fully developed periodically.

The experimental work on mammary cancer in the rat shows that careful, thoughtful experiments, using a model that mimics human breast cancer, always produce meaningful, practical results. Unfortunately, such work in contemporary breast cancer research is the exception rather than the rule.

Independent of pregnancy, women who have had their ovaries removed early in life rarely have breast cancer.[6] Breast cancer risk is also reduced when menarche begins later in life or when menopause is early.[7] These epidemiological studies of risk factors indicate that reduced exposure to estrogen and progesterone lowers the risk of breast cancer. Blocking the effects of estrogen as a protection against breast cancer in healthy women is now being explored in clinical trials.[8]

## AGING

Human experience tells us that cancer tends to be a disease of middle age and older. It also tells us that sunlight can cause premature aging of skin and an increased incidence of skin cancer. As I have noted, cancer and aging are closely linked. (The reason for this association is discussed in Chapter Six.) Aging is a result of the normal functioning, and cancer a result of the defective functioning, of cellular developmental processes in organs.

The skin of a 60-year-old person does not look like the skin of a 20-year-old. However, the skin of a person of 50 can look like the skin of a 60-year-old if he or she is a sun worshipper.

Sunlight destroys skin cells. To replace them, stem cells of the skin, called basal cells, must divide more frequently in the sunbather than in the person who tends to stay out of the sun. One result could be basal cells in the skin of the 50- and 60-year-old of the same biological age because they had gone through similar numbers of mitoses (divisions) during their lives. With each mitosis of a basal cell, the effects of wear and tear on it and the skin grow larger. The tan 50-year-old and the pale 60-year-old have similarly aged skin and share a greater risk of developing skin cancer sometime during their life than a 50-year-old who stays out of the sun.

It seems reasonable to believe, therefore, that we can reduce our risk of cancer, and at the same time slow aging, by reducing the frequency of stem cell doublings in our organs. Keeping our differentiated, mature cells alive longer will reduce the need to replace them, and stem cells will respond by slowing their divisions, keeping them, our organs, and us younger longer. Reducing the demand for mature cells also means that fewer cells will be moving on developmental pathways in organs to be struck by cancer. To benefit, we just must reduce exposure to toxic agents, such as the sun and tobacco, which damage cells.

Even if the total number of our stem cell doublings is genetically determined, the human life span can probably continue to increase. In the 1960s, high jumpers were approaching seven feet; in the 1990s they are at eight feet. Someday they will be close to nine, because of training methods, life style, and desire. Human potential in many areas may be close to unlimited. The less frequent the doubling of stem cells in our organs, the longer our organs may work efficiently.

A factor in the death of elderly people could be that stem cells, within local areas of an organ, have used up their allotment of divisions. As a result, the organ stops functioning properly because repairs cannot be made. The fact that old age is characterized by a diminished number of cells in various organs supports this view.

## THE ENVIRONMENT

The pollution found in large cities damages and kills the mature, differentiated cells of various organs. Higher cell death rates require the stem cells in the affected organs to speed up their doublings, which should be associated with earlier aging. An increased demand for cell replacements also means that more cancer targets—cells in the process of developing—would be present at any time in the affected organs. A greater risk of cancer should result. It would not be a surprise to discover that rural Americans have a significantly longer life expectancy and a lower incidence of cancer than big-city dwellers.

The underlying theme of the preceding material in this chapter is the necessity of looking for ways to regulate cellular developmental processes in organs in order to control malignancy. Elsewhere in the book the same theme is presented. For example, the fact that many breast and prostate tumors are dependent to some extent on the same sex hormones that are essential for the normal development of the breasts and prostate has led to beneficial hormonal therapy for breast and prostate cancer. My work on metastasis, discussed in Chapter Three, suggests that the metastatic process might be stopped if the surfaces of malignant cells were made youthfully "sticky" by the addition of sugars. Drugs or procedures that could return the intracellular biochemical machinery that adds sugars to proteins to its youthful vigor might tame the wandering nature of malignant cells. Since cellular developmental processes in adult organs are always associated with cell division, the unregulated mitoses of malignancy are another likely point at which to try to control cancer.

## VIRUSES

In 1983, while working at the University of North Carolina, I came upon a 1974 paper by Dr. Teruo Asada of Osaka University, Osaka, Japan, in the journal *Cancer* that revealed an interesting perspective on the potential uses of viruses in the treatment of cancer. He wrote, "I assumed that mumps virus would have an affinity and a destructive power to vigorously multiplying cells, such as cancer cells."⁹ Asada's assumption has a logical basis: It is well known that mumps in adult men can cause sterility because the virus kills all of the developing sperm cells. This means that the mumps virus infects and destroys rapidly dividing cells.

Asada described the results of administering live mumps virus to ninety terminally ill patients with various types of cancer. He got very good results in thirty-seven patients and good results in forty-two. Eleven patients were excluded because they were near death. Case reports were presented for ten patients. Following treatment, metastases of a breast tumor to skin "disappeared." In a patient with carcinoma of the gum, the tumor "nearly completely disappeared." The condition of a woman with advanced uterine cancer "improved progressively" after treatment. The case reports went on and on. The results were almost too good to be true and required confirmation before they could be taken seriously.

In medicine, initial reports of impressive advances are always checked out. Procedures are repeated and results duplicated before they are believed and put into general use. If results cannot be confirmed, they are considered spurious.

Although it is possible that confirmatory studies have been conducted since the initial 1974 report, it is more likely that they have not. Even though chemotherapy and immunotherapy are ineffective, they are sanctioned by medical orthodoxy, and oncologists are not likely to stray from convention, even with their terminally ill patients, to try something radically different.

Pharmaceutical and biotechnology companies would also not be interested because anything found in nature, such as a virus, cannot be patented. Without patent protection, there can be competition in the marketplace, and this tends to moderate prices.

Also, unlike the many millions of research and development dollars that are spent to produce proprietary drugs and whose costs are passed along to the patient, growing viruses is easy and inexpensive. For example, the laboratory-grown viruses that are used to make the vaccines that protect against childhood diseases cost drug companies pennies per inoculation to produce. Cheap cancer drugs offer little incentive for companies because inexpensive products are less profitable than expensive ones. A shot that costs pennies to produce and then sells at some higher price produces less profit than a pill with higher production costs and the same percentage of markup. It is really no surprise, therefore, that I have not heard or seen reference to Asada's work or to the viral treatment of cancer.

Let us assume for the sake of discussion that the results are reliable. Asada reported that the destroying action of the mumps virus on cancer cells ended because most adults had acquired an immunity to the virus sometime during their lives. This suggests that therapeutic benefits might be improved by suppressing the immune system before administering the virus. Doses of immunosuppressive drugs would be lowered to control a mumps infection when necessary. Also, a modified virus that is less virulent might prove equally effective while lowering the risk of serious side effects in immunosuppressed patients. What irony if this treatment plan was found to be effective, since cancer doctrine for more than thirty years has stated the complete opposite: Strengthen the immune system to fight cancer. Mavericks who follow the stock market know that when sentiment is all in one direction, benefits can often be had by investing in the other. Why not the same for cancer?

A mumps infection is the likely diagnosis when the pair of salivary glands below and in front of the ears are inflamed. An infection can also be more generalized and can involve the nervous system, testes, prostate, ovaries, breasts, thymus, pancreas, liver, spleen, kidneys, and thyroid gland. Mumps virus has an affinity for the cells of these organs, suggesting that patients with cancers originating there should be used if Asada's idea were to be tested again.

Asada also pointed out that there are a number of other viruses besides mumps that infect and kill rapidly dividing cells. He suggested a kind of combination viral therapy for cancer, in which one kind of virus after another would be administered to maximize the effect. It is an interesting idea and a radical shift away from conventional medicine's reliance on therapies that do not work. I will continue to think that natural viral therapy for cancer has possibilities until additional data from good models are produced that indicate it really does not work.

## CONCLUSIONS

When President Richard Nixon declared the national "war on cancer" in 1971, Congress agreed and appropriated greatly increased funds for the NCI. Their enthusiasm was inspired by our success in space. Only two years before, men had walked on the moon. The NCI predicted that technology would triumph over cancer as well.

Between October 1971 and March 1972, 250 basic and clinical scientists met in a series of forty planning sessions and two major review sessions.[10] Their goal was to develop a scientific plan for the National Cancer Program. The most important objective of the final plan was to develop the means to cure patients with cancer. More than twenty years later, while technology solves the problems of space modules millions of miles away, the National Cancer Program has yet to provide any answers for the millions of people who are struck with cancer here on earth. Scientific understanding from laboratory research has not paved the way to practical applications.

If we are ever to win the war against cancer, the fundamental balance of power in cancer research must be radically shifted. At present, the cancer research establishment is nearly devoid of people who understand the characteristics of human cancer. Concerned pathologists must try to open up the insular worlds of cancer research and molecular biology to the light of reason. Pathology must try to reestablish its traditional role in cancer research as the final judge of the suitability of experimental models, so that cell lines can be discredited by a powerful and respected voice in medicine.

As public health issues such as breast cancer and AIDS have heated up, concerned patients and potential patients have become increasingly vocal in their demands for answers. What causes these diseases? How can we prevent them? How can they be cured?

However, what few activists realize is that it is not enough simply to demand more money for research. As long as researchers continue to search for truth in petri dishes, valid treatment options will never be discovered. It is time to demand a radical shift in the cancer paradigm—the abandonment of cell lines and a return to models that more closely approximate reality.

Cancer should concern all of us. A concerned group in the gay community, ACT UP, has become a force to reckon with in only a few years. This should teach each of us that it is very difficult to do something by oneself; we must organize to bring about change. In the United States, the number of deaths from breast cancer alone in the 1980s was roughly six times the American fatalities from AIDS.[11]

Throwing more money into cancer research at the present time is not the answer, and perhaps for many of the same reasons, more money is also not the answer for AIDS (see AIDS, page 176). Additional funding would actually make the cancer situation *worse* since a large and powerful bureaucracy looking into a petri dish for answers would become even more firmly mired in the scum.

Activists can achieve their goals by organizing and making

# AIDS

Many AIDS activists are concerned that the battle against AIDS will become another war on cancer. Their concern is well founded. The federal medical and scientific bureaucracy that has wasted billions of dollars on cancer also directs the research on AIDS. In the July 1992 issue of "Bio/Technology" magazine, a noted AIDS researcher commented on the national research program on AIDS and the development of an AIDS vaccine. Speaking on condition of anonymity, he said,

> Until the immunology and pathogenesis of various diseases are more fully understood, vaccine research will not have a solid foundation on which to base any vaccine.[12]

AIDS researchers, like their colleagues in the cancer field, do not have a basic understanding of a disease or of the virus that is believed to cause it. This means that AIDS researchers also commonly attribute experimental findings to a disease that are really irrelevant to it. The prominent AIDS researcher continued,

> If anyone dares voice a minority view it is often shouted down. As a result new concepts in AIDS research originating from less well-known investigators gain acceptance more slowly . . . . If dogmatists push only one experimental design—what kind of people are they training? . . . Prima donnas get all the press, publications, meeting invitations, more access to grant money and the right to peer review their younger colleagues. This implies the power to destroy budding careers that don't follow the mainstream.[13]

The AIDS story is all too familiar, but why should it be any different? The war on AIDS has been modeled on the war on cancer.

their specific concerns known to Congress. Until such action is taken, cancer patients will continue to be subjected to toxic and ineffective treatments and researchers will continue to waste billions of tax and private research dollars. We need to learn that cancer researchers have an extraordinary capacity for self-delusion. As a result, they have been misleading themselves and the rest of society for a very long time. America is facing a failing health care system. A significant part of our health research system is failing as well. A hard look must be taken at the cancer industry because behind thousands of laboratory doors is the biomedical equivalent of the savings and loan debacle—and we are all paying the price for it.

# Postscript

During the spring of 1993, shortly after *The Immortal Cell* had been turned over to the publisher for typesetting, two new cancer research failures broke in the media. Not surprisingly, these two related events matched a gloomy prediction I had already made in the manuscript.

First, in March, one of the biggest players in the cancer war, Dr. Steven Rosenberg (see page 124), was rebuffed by a scientific advisory committee of the National Institutes of Health. The committee, upon review of Rosenberg's research results, determined that the technology that Rosenberg was bringing to bear on the cancer problem was not working.

Three years before, Rosenberg had claimed that his work had demonstrated that gene therapy—introducing new genes into the cells of cancer patients—had promise as a treatment for cancer. He had been given in excess of $1 million to launch clinical trials of this new breakthrough.[1]

The experiment entailed injecting the gene for tumor necrosis factor (TNF) into lymphocytes harvested from cancer patients and grown in culture. These modified lymphocytes were then reinjected into the patients. Rosenberg expected the lymphocytes to home in on the tumors and secrete TNF, killing the cancer cells. He was wrong.

After three years of clinical trials, it was clear that Rosenberg's great breakthrough was a flop. The committee unanimously voted to delete a quarter of a million dollars from Rosenberg's annual $1 million-plus research budget.[2]

The second great blow occurred in April, when the April 30 issue of *The Wall Street Journal* announced that Genentech Inc. (America's first biotechnology company) had removed tumor necrosis factor from its list of research projects. Genentech had worked on TNF for a decade (see Chapter Twelve), betting millions of dollars that the protein would kill tumor cells in cancer patients.[3] They should have gone short on TNF—as I did.

With each successive research failure, it becomes more and more clear that current gene therapy approaches will never yield the cancer treatments we so desperately need. Existing research cannot hope to bridge the gap between the petri dish and the cancer patient, because it is based on an inherently unsound premise—that cell lines and human cancer cells are one and the same.

The demise of tumor necrosis factor, like that of interleukin 2 and interferon before it, underscores the tremendous importance of abandoning incorrect experimental research models. Only when research reflects an intimate knowledge of and feeling for cancer as it occurs in the dynamic environment of the body—made possible by years of close association with the real "organism"—will we be able to make real progress in the war on cancer.

# APPENDIX

# The Human Genome Project

*Despite its glamour, DNA is simply the construction manual that directs the assembly of the cell's proteins. The DNA itself is lifeless, its language cold and austere.*[1]

Arthur Kornberg, M.D.,
Nobel Laureate in Medicine
*For the Love of Enzymes,* 1989

Cell lines and the methods of molecular biology have had an impact that reaches far beyond the boundaries of cancer research. The concept that humans as a whole—like individual cancer cells—can be reduced to their genes is the driving force behind one of the most massive scientific research projects of all time, the Human Genome Project (HGP).

In 1990, the National Institutes of Health and the Department of Energy launched the $3-billion, fifteen-year project. The human genome is the genetic information that is transmitted to every child. The project's ambitious goal is to decode this chemical blueprint for human life. Using the methods of molecular biology, project researchers intend to produce an accurate map of the locations of all the genes in our forty-six chromosomes, as well as the complete sequence of the 3 billion subunits of chromosomal DNA that make up the genetic code.[2] Systematically sequencing the genome is equivalent to arranging all the letters in about ten sets of the Encyclopedia Britannica.

Dr. Daniel E. Koshland, Jr., the editor of Science magazine, supports the HGP. In 1989, he said that "sequencing the human genome puts us on the threshold of great new benefits."[3] One of these "great new benefits" is supposed to be early-warning tests for cancer. Koshland's remark is reminiscent of the cancer experts who claim that molecular biology's advanced technology puts us on the threshold of conquering cancer. Should we believe Koshland any more than the others?

Genetic tests for cancer would be a reasonable expectation of the HGP, for most of the population, if people with cancer usually had a strong family history of it. But most people do not have this risk factor. The relationship of genes to most cancers is indirect at best. Scientists such as Koshland, who think the HGP will lead to cancer warning tests, are assuming not only that cancer is caused by mutated genes but also that a cell must accumulate a number of defects before it becomes malignant. As was discussed earlier, both of these assumptions may be false.

According to Nobel laureate Salvador Luria of MIT, the project's approval was the result of intense lobbying of Congress and the scientific community "by a small coterie of power seeking enthusiasts."[4] In The New York Times on June 5, 1990, Dr. Martin Rechsteiner, a biochemist at the University of Utah, was quoted as saying that the HGP is "bad science, it is hyped science." In the same article, Dr. Michael Syvanen, a microbiologist at the medical school of the University of California, Davis, said that "everybody I talk to thinks this is an incredibly bad idea."[5]

But why is the Human Genome Project bad science and a bad idea? The DNA of genes is the information of life, the instruction books kept in a central library. When the instruction books are read by cellular machinery, proteins are made and the activities of life follow naturally. Proponents of the HGP would like everyone to believe that human life can be reduced to this information within the chemical language of DNA. To say that humans reduce to their genes is equivalent to saying that a major league baseball game reduces to the lineup cards of the managers at the start of the game. These cards, although useful information, can tell us nothing about the actual play of the game—the drama, the hits, the errors, the stolen bases, and so on. Each game is characterized by an ebb and flow of play that is unpredictable and unique. In contrast, the lineup cards are static, except for the removal and addition of players.

The goal of the HGP is to provide the lineup of the human cell. Although it will define the players, the proteins, it will not tell us anything about the game itself, the dynamic processes of human life. These processes involve complex interactions among thousands of proteins that occur mainly in a cell's cytoplasm (the playing field), not in the nucleus (the dugout) where the chromosomes containing the genes are stored.

Gene maps tell us nothing about how the thousands of proteins of the cell work together to produce life. The processes of life are not static, although molecular biologists would have us believe that they are. Determining the precise order of the 3 billion subunit pairs of DNA will provide very detailed, very media-worthy, very expensive information with very little practical use. Although the lineup card of life will be filled, we still will not know how the game itself is played.

Researchers at the University of Colorado, Boulder, have determined that the total number of different proteins in a human cell is in the range of 4,000.[6] Of these 4,000 proteins, perhaps 80 percent, or 3,200, perform basic functions common to all the cells of the body. The specialized characteristics of each cell type, such as muscle cell contractions or nerve cell electrical impulses, would be determined by the remaining 20 percent, or 800 proteins. Since there are about 200 different cell types in a human, each with 3,200 common proteins and 800 unique proteins, there may be only about 20,000 genes for

20,000 different proteins in a human genome of 46 chromosomes, large enough to contain about a million genes.

To spend $3 billion to confirm that only 2 percent of human DNA contains genes would be a very poor investment. In other words, 98 percent of our DNA may have no other function than to carry 20,000 essential genes that take up only 2 percent of the space. The fact that the bone marrow cells and lymphocytes of atomic bomb survivors functioned normally with many chromosomal mutations (see page 84) indicates that most of our DNA may indeed be "junk."

But the biggest complaint against the HGP has nothing to do with its objectives. Rather, it has to do with the human cells from which the project intends to derive the lineup card of life. In February 1990, I contacted Dr. John Kelsoe of the University of California, San Diego, to discuss his research, which dealt with the gene that causes a form of manic depressive disease. During our telephone conversation, he told me that the plan of the HGP designates cell lines as the source of chromosomes for gene mapping. An information officer at the National Institutes of Health's Genome Center in Bethesda, Maryland, later confirmed Kelsoe's statement. Given their known, pronounced chromosomal instability, what are the chances that cell lines will provide an accurate gene map?

In tracking down the genes that cause inherited diseases, such as muscular dystrophy, cystic fibrosis, Huntington's disease, or a form of manic depression, cell lines are usually established from the blood cells of individuals with a family history of the disease. However, there have been problems with this approach. Sometimes one group cannot confirm the results of another. Contradictory data have pointed to different chromosomal locations for the same gene. In 1987, a much-ballyhooed paper linked manic depressive disease to a place in chromosome 11.[7] By the end of 1989, Kelsoe was reporting to his colleagues that he could not find the gene there.[8]

Those involved in this work should consider the very real

possibility that their experimental model had tricked them. Like all cell lines, those that are used to locate the gene for manic depressive disease have unstable chromosomes. They break and rejoin in unnatural ways, forming chromosomal rearrangements, duplications, and deletions. This means that two laboratories could find the manic depressive gene at different chromosomal locations and both labs could be wrong. Confusion of this sort will be multiplied many times over by the time the HGP is supposed to be completed.

# Notes

## Chapter 1
### Losing Ground

1. F.M. Burnet, *Genes, Dreams, and Realities* (New York: Basic Books, 1971), 132.
2. J.C. Bailar and E.M. Smith, "Progress Against Cancer?" *New England Journal of Medicine* 314 (1986), 1226–1232.
3. Media Advisory, Fenton Communications, February 4, 1992, Washington, D.C.
4. American Cancer Society, *Cancer Facts & Figures—1991* (Atlanta: American Cancer Society, 1991).
5. Provisions of National Cancer Act of 1937, Foreword, *Journal of the National Cancer Institute* 1 (1940): I.
6. Burnet, *Genes, Dreams, and Realities*, 132.
7. Cited in S. Weinhouse, "National Cancer Act of 1971—An Editorial," *Cancer Res.* 32 (1972), i–ii.

8. Bailar and Smith, "Progress Against Cancer?"
9. M.J. Wizenberg, "Rocket Research and Pony Express Delivery," *Cancer* 63 (1989), 2387–2392.
10. B.R. Cassileth, E.J. Lusk, D. Guerry, et al., "Survival and the Quality of Life Among Patients Receiving Unproven as Compared with Conventional Cancer Therapy," *New England Journal of Medicine* 324 (1991), 1180–1185.
11. Cited in M. Snider, "Treating Incurable Cancer May Not Extend Life," *USA Today*, 25 April 1991, 1.

## Chapter 2
### The Making of a Cancer Scientist

1. I.M. Arias, "Training Basic Scientists to Bridge the Gap Between Basic Science and Its Application to Human Disease," *New England Journal of Medicine* 321 (1989), 972–973.

**Chapter 3**
**The Road to Dissent**

1. *The Life and Selected Writings of Thomas Jefferson* (New York: Random House, 1972), 201.
2. Cited in L. Balter, "Three Europeans Find Their Own Road to Fame," *Science* 254 (1991), 1116–1117.
3. G.B. Dermer, "Changes in the Surface Coats of Neoplastic Human Breast Epithelium," *Cancer Res.* 33 (1973), 999–1002.
4. G.B. Dermer and W.H. Kern, "Changes in the Affinity of Phosphotungstic Acid and Positively Charged Colloidal Particles for the Surfaces of Malignant Human Transitional Epithelium of the Urinary Bladder," *Cancer Res.* 34 (1974), 2011–2014.
5. J. Burchell and J. Taylor-Papadimitriou, "Antibodies to Human Milk Fat Globule Molecules," *Cancer Invest.* 7 (1989), 53–61.
6. G.B. Dermer and R.P. Sherwin, "Autoradiographic Localization of Glycoprotein in Human Breast Cancer Cells Maintained in Organic Culture After Incubation with Tritiated Fucose or Glucosamine," *Cancer Res.* 35 (1975), 63–67.
7. G.B. Dermer, "Autoradiology of Cellular Glycoproteins Reveals Histogenesis of Bronchogenic Adenocarcinoma," *Cancer* 47 (1981), 2000–2006.

8. Ibid.
9. *The Life and Selected Writings of Thomas Jefferson*, 209.
10. G.B. Dermer, "Human Cancer Research," *Science* 221 (1983), 318.

**Chapter 4**
**The Immortal Cell**

1. J.J. Clausen and J.T. Syverton, "Comparative Study of 31 Cultured Mammalian Cell Lines," *J. Nat. Cancer Inst.* 28 (1961), 117–131.
2. W.F. Schere, J.T. Syverton, and G.O. Gey, "Studies on the Propagation *In Vitro* of Poliomyelitis Viruses VI. Viral Multiplication in a Stable Strain of Human Malignant Epithelial Cells (Strain HeLa) Derived From an Epidermoid Carcinoma of the Cervix," *Journal of Experimental Medicine*, 97 (1953), 695–710.
3. L.A. Loeb, "Endogenous Carcinogenesis: Molecular Oncology into the Twenty-First Century," *Cancer Research*, 49 (1989), 5489–5496.
4. Ibid.

**Chapter 5**
**The Contradictions of Stability and Differentiation**

1. J.C.A. Recamier, "Recherches sur le traitment du cancer, par la compression méthodique simple oú combinée, et sur

l'histoire générale de le même maladie," Vol. 2 (Paris: Gabon, 1829), 110.

2. Cited in L.J. Rather, *The Genesis of Cancer: A Study in the History of Ideas* (Baltimore: Johns Hopkins University Press, 1978), 166.

3. W. Moxon, Discussion on Cancer, Transactions of the Pathology Society, London, 25 (1874): 346.

4. E.D. Day, "Animal Cancer, the Primary Tool," *Cancer Research* 21 (1961), 581–582.

5. N.B. Atkin, "Solid Tumor Cytogenetics," *Cancer Genet. Cytogenet.* 40 (1989), 3–12.

6. Ibid.

7. P.J. Whang, E.C. Lee, C. Kao-Shan, et al., "Cytogenetic Studies of Human Breast Cancer Lines: MCF-7 and Derived Variant Sublines," *Journal of the National Cancer Institute* 71 (1983), 687–691.

8. Ibid.

9. J. Fogh, "Human Tumor Lines for Cancer Research," *Cancer Invest.* 4 (1986), 157–184.

10. E.H.Y. Chu and H. Giles, "Comparative Chromosomal Studies on Mammalian Cells in Culture. I. The HeLa Strain and Its Mutant Clonal Derivatives," *Journal of the National Cancer Institute* 20 (1958), 383–395.

11. A.E. Moore, C.M. Southam, and S.S. Sternberg, "Neoplastic Changes Developing in Epi-thelial Cell Lines Derived From Normal Persons," *Science* 124 (1956), 127–129.

12. Chu and Giles, "Comparative Chromosomal Studies on Mammalian Cells in Culture."

13. K.H. Walen and M.R. Stampfer, "Chromosome Analyses of Human Mammary Epithelial Cells at Stages of Chemical-Induced Transformation Progression to Immortality," *Cancer Genet. Cytogenet.* 37 (1989), 249–261.

14. Ibid.

15. Cited in transcript of *Proceedings, President's Cancer Panel Meeting,* December 7, 1990, National Cancer Institute (Silver Spring, MD: Eberlin Reporting Service, 1990), p. 90.

16. M. Gold, *A Conspiracy of Cells* (Buffalo: University of New York Press, 1986).

17. W. Nelson-Rees, R.F. Flandermeyer, and P.K. Hawthorne, "Banded Marker Chromosomes as Indicators of Intraspecies Cellular Combination," *Science* 184 (1974), 1093–1096.

## Chapter 6
### The Contradictions of Initiation and Metastasis

1. Cited in L.J. Rather, *The Genesis of Cancer: A Study in the History of Ideas* (Baltimore: Johns Hopkins University Press, 1978), 152.

2. S. Weinhouse, "Glycosis, Respiration, and Anomalous Gene

Expression in Experimental Hepatomas," *Cancer Research* 32 (1972), 2007–2016.

3. G.B. Dermer, "Basal Cell Proliferation in Benign Prostatic Hyperplasia," *Cancer* 41 (1978), 1857–1862.

4. K.M. Pozharisski, "Morphology and Morphogenesis of Experimental Epithelial Tumors of the Intestine," *Journal of the National Cancer Institute* 54 (1975), 1115–1123.

5. Ibid.

6. H. Varmus and R.A. Weinberg, *Genes and the Biology of Cancer* (New York: Scientific American Library, 1993), 79–82.

7. J.M. Bishop, "The Molecular Genetics of Cancer," *Science* 235 (1987), 305–311.

8. P.H. Duesberg, "Cancer Genes: Rare Recombinants Instead of Activated Oncogenes," *Proc. Natl. Acad. Sci. USA* 84 (1987), 2117–2124.

9. I.J. Fidler, "Selection of Successive Tumor Lines for Metastasis," *Nature* 242 (1973), 148–149.

10. G. Yogeeswaran and P. Salk, "Metastatic Potential is Positively Correlated with Cell Surface Sialylation of Cultured Murine Tumor Cell Lines," *Science* 212 (1981), 1514–1516.

11. S. Trinidad, J.R. Lisa, and M.B. Rosenblatt, "Bronchogenic Carcinoma Simulated by Metastatic Tumors," *Cancer* 16 (1963), 1521–1529.

**Chapter 7**
**The Elusive Oncogene**

1. P.H. Duesberg, "Activated Proto-Oncogenes: Sufficient or Necessary for Cancer," *Science* 228 (1985), 669–677.

2. R. Sager, "Tumor Suppressor Genes: The Puzzle and the Promise," *Science* 246 (1989), 1406–1411.

3. R.J. Huebner and G.J. Todaro, "Oncogenes of RNA Tumor Viruses as Determinants of Cancer," *Proc. Natl. Acad. Sci. USA* 64 (1969), 1087–1094.

4. M.P. Calos, J.S. Lebkowski, and M.R. Botchan, "High Mutation Frequency in DNA Transfected Into Mammalian Cells," *Proc. Natl. Acad. Sci. USA* 80 (1983), 3015–3019, and A. Razzaque, H. Mizusawa, and M.M. Seidman, "Rearrangement and Mutagenesis of a Shuttle Vector Plasmoid After Passage in Mammalian Cells," *Proc. Natl. Acad. Sci. USA* 80 (1983), 3010–3014.

5. F. Mitelman, Y. Kaneka, and J.M. Trent, "Report of the Committee on Chromosome Changes in Neoplasia," *Cytogenet. Cell Genet.* 55 (1990), 358–386.

6. L.A. Loeb, "Mutator Phenotype May be Required for Multistage Carcinogenesis," *Cancer Research* 51 (1991), 3075–3079.

7. P.G. Chesa, W.G. Rettig, M.R.

Melamed, et al., "Expression of p21-*ras* in Normal and Malignant Human Tissues. Lack of Association with Proliferation and Malignancy," *Proc. Natl. Acad. Sci. USA* 84 (1987), 3234–3238.

Chapter 8
*The Epigenetic Theory*

1. I.B. Weinstein, "A Possible Role for Transfer RNA in the Mechanism of Carcinogenesis," *Cancer Research* 28 (1968), 1871–1874.
2. S.I. Hajdu and J. Fogh, "The Nude Mouse as a Diagnostic Tool in Human Tumor Cell Research," *The Nude Mouse in Experimental and Clinical Research*, ed. J. Fogh and B.C. Giovanilli (New York: Academic Press, 1979), 235–266.
3. V.L. Wilson, R.A. Smith, S. Ma, et al., "Genomic 5-Methyldeoxycytidine Decreases With Age," *J. Biol. Chem.* 262 (1987), 9948–9951; S.W. Sherwood, D. Rush, J.L. Ellsworth, et al., "Defining Cellular Senescene in IMR-90 Cells: A Flow Cytometric Analysis," *Proceedings of the National Academy of Sciences USA* 85 (1988), 9086–9090; and C.G. Harley, A.B. Futcher, and C.W. Credier, "Telomeres Shorten During Aging of Human Fibroblasts," *Nature* 345 (1990), 458–460.
4. Y. Tomonaga, "Chromosome Abnormalities in Atomic Bomb Survivors," *Nagasaki Med. J.* 51 (1976), 282–286.
5. H.C. Pitot and C. Heidelberger, "Metabolic Regulatory Circuits and Carcinogenesis," *Cancer Research* 23 (1963), 1694–1700.

Chapter 9
*The Politics of Cancer: "The Bottom Line Is Dollars"*

1. D.M. Cooper, "The Human Genome Program," *Science* 246 (1989), 874.
2. B. Nussbaum, *Good Intentions* (New York: Atlantic Monthly Press, 1990), 332.
3. A. Chase, "Politics and Science," *Orange County Register*, 6 October 1991, J1 J2.
4. *Cancer Facts & Figures—1991* (Atlanta: American Cancer Society, 1991).
5. Membership roster, National Cancer Advisory Board, 1989.
6. Ibid.
7. Membership roster, National Cancer Panel, 1989.
8. National Cancer Institute, *NCI Fact Book*, 1990, 65–66.
9. Cited in L. Roberts, "News and Comment," *Science* 246 (1989), 204–205.
10. Arizona Cancer Center, *Annual Report (1988)* (Tucson: University of Arizona, 1988).
11. Ibid, 2.
12. Ibid, 4.

13. A.I. Holleb, "Progress Against Cancer? A Broader View," *CA— A Cancer Journal for Clinicians* 36 (1986), 243–244.

14. *CBS Evening News*, 16 December 1991, Transcript (Burrelle's Information Services), 6.

15. D.S. Greenberg, "A Critical Look at Cancer Coverage," *Columbia Journalism Rev.*, January/February 1975, 40–44.

16. Ibid.

17. Ibid.

18. Ibid.

19. Ibid.

**Chapter 10**
*Chemotherapy: The Great Hope*

1. K. Endicott, "The Chemotherapy Program," *Journal of the National Cancer Institute* 19 (1957), 275–285.

2. D.E. Young, "An Ethical Approach to Chemotherapy in Private Practice," *Journal of the National Cancer Institute* 84 (1992), 810.

3. A. Gilman and F.S. Philips, "The Biological Actions and Therapeutic Applications of the β-Chlorethyl Amines and Sulfides," *Science* 103 (1946), 409–415.

4. J.F. Holland and C. Heidelberger, "Human Cancer, the Primary Target," *Cancer Research* 20 (1960), 975–976.

5. M.R. Boyd, "Status of the Implementation of the NCI Human Tumor Cell Line *in Vitro* Primary Drug Screen," *Proceedings of the American Association for Cancer Research* 30 (1989), 652–654.

6. L.M. Weisenthal, "Antineoplastic Drug Screening Belongs in the Laboratory, Not in the Clinic," *Journal of the National Cancer Institute* 84 (1992), 466–469.

7. S.F. Stinson, M.C. Alley, S. Kenny, et al., "Morphologic Characterization of Human Carcinoma Cell Lines," *Proceedings of the American Association for Cancer Research* 30 (1989), 613.

8. T.H. Corbett, L. Polin, A.J. Wozniak, et al., "The Use of a Multiple-Tumor-Soft-Agar-Disk-Diffusion-Assay to Detect Agents with Selective Solid Tumor Activity," *Proceedings of the American Association for Cancer Research* 29 (1988), 533–534.

9. R. Teitelman, *Gene Dreams: Wall Street, Academia, and the Rise of Biotechnology* (New York: Basic Books, 1989), 145.

10. G.A. Omura, M.F. Brady, H.D. Homesley, et al., "Long-Term Follow-Up and Prognostic Factor Analysis in Advanced Ovarian Cancer: The Gynecologic Oncology Group Experience," *Journal of Clinical Oncology* 9 (1991), 1138–1150.

11. P.C. Pappas, et al., "Marked Discrepancy Between Survival

and Tumor Response End-points in Cancer Trials," *Proceedings of the American Society of Clinical Oncology* 11 (1992), 156.

12. M.J. Kennedy, R. Beveridge, S. Rowley, et al., "Dose-Intense Cytoreduction Followed by High Dose Consolidation Chemotherapy and Rescue with Purged Autologous Bone Marrow for Metastatic Breast Cancer," *Program, 12th Annual San Antonio Breast Cancer Symposium: In Breast Cancer Research and Treatment* 14 (1989), 110.

13. D. Young, Letter to the editor, *Journal of the National Cancer Institute* 84 (1992), 1533.

14. House Select Committee on Aging, 98th Congress, May 31, 1984, *Quackery: A $10 Billion Scandal* (Washington, DC: U.S. Government Printing Office, 1984).

## Chapter 11
### *Immunotherapy: The Great Fad*

1. H.K. Hewitt, Letter to the editor, *Cancer Research* 39 (1979), 4285–4287.

2. C.G. Moertel, "On Lymphokines, Cytokines, and Breakthroughs," *Journal of the American Medical Association* 256 (1986), 3117.

3. A. Marchetti, *Beating the Odds: Alternative Treatments That Have Worked Miracles Against Cancer* (Chicago: Contemporary Books, 1988), 171–172.

4. D. Guerry and L.M. Schuchter, "Disseminated Melanoma—Is There a New Standard Therapy?" *New England Journal of Medicine* 327 (1992), 561–562.

5. P. Alexander, "Back to the Drawing Board—The Need for More Realistic Systems for Immunotherapy," *Cancer* 40 (1977), 467–470.

6. O.C. Scott, "Tumor Transplantation and Tumor Immunity: A Personal View," *Cancer Research* 51 (1991), 757–763.

7. D.J. Fitzgerald, J.W. Pearson, D.L. Longo, et al., "Antitumor Activity of OVB3-PE Immunotoxin in Nude Mice Bearing Ovarian and Colon Tumors," *Proceedings of the American Association for Cancer Research* 30 (1989), 646–647.

8. J.J. Mule, J.C. Yang, R.L. Afrenier, et al., "Identification of Cellular Mechanisms Operational *in Vivo* During the Regression of Established Pulmonary Metastases by the Systematic Administration of High-Dose Recombinant Interleukin 2," *Journal of Immunology* 139 (1987), 285–294.

9. Discussed in B.J. Culliton, "Gene Test Begins," *Science* 244 (1989): 913.

10. M.S. McCabe, D. Stablein, and M.J. Hawkins, "The Modified Group C Experience: Phase III Randomized Trials of IL-2 vs. IL-2/LAK in Advanced Renal

Cell Carcinoma and Advanced Melanoma," *Proceedings of the American Society of Clinical Oncology* 10 (1991), 213.

11. H.C. Hoover, M.G. Surdyke, R.B. Dangel, et al., "Prospectively Randomized Trial of Adjuvant Active-Specific Immunotherapy for Human Colorectal Cancer," *Cancer* 55 (1985): 1236–1243.

12. *Nineteenth Report by the Committee on Government Operations of the House of Representatives. Are Scientific Misconduct and Conflicts of Interest Hazardous to Our Health*, 101st Congress, House Report 101-688 (10 September 1990).

13. Ibid, 149.

14. "Michael Landon Mourned," *San Antonio Express–News*, 2 July 1991, 1D, 5D.

15. "Cancer Claims Lee Remick," *Arizona Republic*, 3 July 1991, A1, A17.

**Chapter 12**
*Unmasking Biotechnology*

1. M. Crichton, *Jurassic Park* (New York: Ballantine Books, 1991), x.

2. D. Gilbert, "Commentary on Wall Street," *Bio/Technology* 10 (1992), 1518–1519.

3. D.S. Greenberg, "Health-Care Spending—Up, Up, and Away," *Lancet* 340 (1992), 1086–1087.

4. C.G. Moertel, "Off-Label Drug Use for Cancer Therapy and National Health Care Priorities," *Journal of the American Medical Association* 266 (1991), 3031–3032.

5. M. Edelhart, *Interferon, the New Hope for Cancer* (Reading, MA: Addison-Wesley, 1981).

6. K.A. Foon, "Biological Response Modifiers: The New Immunotherapy," *Cancer Research* 49 (1989), 1621–1639.

7. G. Bylinsky, "Science Scores a Cancer Breakthrough," *Fortune*, November 25, 1985, 16–21.

8. Cited in M. Clark, et al., "Search for a Cure," *Newsweek*, 16 December 1985, 60–65.

9. E. Clark and R. Hoppe, "The New War on Cancer," *Business Week*, 22 September 1986, 60–63.

10. Ibid.

11. B.J. Culliton, "Cetus's Costly Stumble on IL-2," *Science* 250 (1990), 20–21.

12. Ibid.

13. M. Ratner, "The Cetus Experience: Troubles with Clinical Trials," *Bio/Technology* 8 (1990), 815–818.

14. B.J. Spalding, "Chiron and Cetus Wed in $660-Million Deal," *Bio/Technology* 9 (1991), 789–790.

15. Ibid.

16. B.J. Spalding, "87 Biopharmaceutical Firms Lose $662 Million," *Bio/Technology* 11 (1993), 426–428.

17. A.L. Sprout, "America's Most Admired Companies," *Fortune*, 11 February 1991, 52–64.

18. Clark and Hoppe, "The New War on Cancer."

19. D. Guerry and L.M. Schuchter, "Disseminated Melanoma—Is There a New Standard Therapy?" *New England Journal of Medicine* 327 (1992), 561–562.

20. F. Crick, *What Mad Pursuit* (New York: Basic Books, 1988), 142.

21. R. Teitelman, *Gene Dreams: Wall Street, Academia and the Rise of Biotechnology* (New York: Basic Books, 1989), 215.

22. E.A. Carswell, L.J. Old, R.L. Kassel, et al., "An Endotoxin-Induced Serum Factor That Causes Necrosis in Tumors," *Proceedings of the National Academy of Science USA* 72 (1975), 3666–3670, and L. Helson, S. Green, E.A. Carswell, et al., "Effect of Tumor Necrosis Factor on Cultured Human Melanoma Cells," *Nature* 258 (1975), 731–732.

**Chapter 13**
*Every Woman's Nightmare*

1. G. Radner, *It's Always Something* (New York: Simon & Schuster, 1989), 85.

2. T. Weiss, Opening statement, Congressional Hearing on Breast Cancer Research and Treatment, 102nd Congress, Committee on Government Operations, Subcommittee on Human Resources and Intergovernmental Relations, 11 December 1991, 1.

3. D. Hutton, "Breast Cancer," *Vogue* (British) 157 (1993), 120–123, 146–147.

4. R.A. Spear, Testimony, Subcommittee on Human Resources and Intergovernmental Relations Hearing on Breast Cancer Research and Treatment, 11 December 1991, 3.

5. Radner, *It's Always Something*, 72.

6. E. Petru and D. Schmahl, "No Relevant Influence on Overall Survival Time in Patients with Metastatic Breast Cancer Undergoing Combination Chemotherapy," *Journal of Cancer Res. Clin. Oncol.* 114: (1988), 183–185.

7. M. Beck, E. Yoffe, G. Carroll, et al., "The Politics of Breast Cancer," *Newsweek*, 10 December 1990, 62–65.

8. Statement of Dr. Bernadine Healy, director, National Institutes of Health, Subcommittee on Human Resources and Intergovernmental Relations Hearing on Breast Cancer Research and Treatment, 11 December 1991, 6.

9. H.D. Soule, J. Vasquez, A. Long, et al., "A Human Cell Line from a Pleural Effusion Derived From a Breast Carcinoma," *Journal of the National*

*Cancer Institute* 51 (1973), 1409–1416.

10. A. Weil, *Health and Healing* (Boston: Houghton Mifflin, 1983).
11. Beck, Yoffe, Carol, et al., "The Politics of Breast Cancer."
12. V. Ling, et al., "Multidrug Resistance Phenotype in Chinese Hamster Ovary Cells," *Cancer Treatment Reports* 67 (1983), 869–874.
13. G. Bonadonna, "Evolving Concepts in the Systemic Adjuvant Treatment of Breast Cancer," *Cancer Research* 52 (1992), 2127–2137.
14. E.L. Wynder, D.P. Rose, and L.A. Cohen, "Diet and Breast Cancer in Causation and Therapy, *Cancer* 58 (1986), 1804–1813.
15. E. Marshall, "Search for a Killer: Focus Shifts from Fat to Hormones," *Science* 259 (1993), 618–621.
16. Ibid.
17. Ibid.
18. J.B. Westin and E. Richter, "The Israeli Breast Cancer Anomaly," *Annals NY Acad. Sci.* 609 (1990), 269–279.
19. Ibid.
20. C.O. Grani, "Ovarian Cancer: Unrealistic Expectations," *New England Journal of Medicine* 327 (1992), 197–199.
21. "Does Health Insurance Deliver on Its Promises?" *Larry King Live*, Transcript #271, 2 April 1991.
22. Ibid., p. 4.
23. Granai, "Ovarian Cancer."

## Chapter 14
### The Real Picture

1. B. Barber, "Resistance by Scientists to Scientific Discovery," *Science* 134 (1961), 596–602.
2. P.H. Duesberg, "Activated Proto-Oncogenes: Sufficient or Necessary for Cancer?" *Science* 228 (1985), 669–677.
3. T.S. Kuhn, *The Structure of Scientific Revolutions*, 2nd ed. (Chicago: University of Chicago Press, 1970).
4. G. Nicolson, "Tumor Cell Instability, Diversification, and Progression of the Metastatic Phenotype: From Oncogene to Oncofetal Expression," *Cancer Research* 47 (1987), 1473–1478.
5. E.H.Y. Chu and H. Giles, "Comparative Chromosomal Studies on Mammalian Cells in Culture. I. The HeLa strain and its Mutant Clonal Derivatives," *Journal of the National Cancer Institute* 20 (1958), 383–395.
6. See, for example, S. Goldstein, "Replicative Senescence: The Human Fibroblast Comes of Age," *Science* 249 (1990), 1129–1132.

## Chapter 15
### The Promise of Correct Models

1. F. Crick, *What Mad Pursuit* (New York: Basic Books, 1988), 161.
2. X. Fu, J.M. Besterman, A. Monosov, et al., "Models of Hu-

man Metastatic Colon Cancer in Nude Mice Orthotopically Constructed by Using Histologically Intact Human Specimens," *Proceedings of the National Academy of Sciences USA* 88 (1991), 9345–9349.

3. Ibid.

4. B. MacMahon, P. Cole, T.M. Lin, et al., "Age at First Birth and Breast Cancer Risk," *Bul. WHO* 43 (1970), 209–221.

5. J. Russo and I.H. Russo, "Modulation of the Mammary Gland's Susceptibility to Carcinogenesis," *Proceedings of the American Association for Cancer Research* 29 (1988), 529–531.

6. E. Marshall, "Search for a Killer: Focus Shifts from Fat to Hormones," *Science* 259 (1993), 618–621.

7. Ibid.

8. B.E. Henderson, R.K. Ross, and M.C. Pike, "Hormonal Chemo-Prevention of Cancer in Women," *Science* 259 (1993), 633–638.

9. T. Asada, "Treatment of Human Cancer with Mumps Virus," *Cancer* 34 (1974), 1907–1928.

10. U.S. Department of Health, Education and Welfare/Public Health Service of the National Institutes of Health, *Digest of Scientific Recommendations for the National Cancer Program Plan*, DHEW publication No. 74–570, 1973, *i–vi*.

11. M. Beck, E. Yoffe, G. Carroll, et al., "The Politics of Breast Cancer," *Newsweek*, 10 December 1990, 62–65.

12. S.M. Edgington, "Is an AIDS Vaccine Possible?" *Bio/Technology* 10 (1992), 768–771.

13. Ibid.

## Postscript

1. C. Anderson, "A Speeding Ticket for NIH's Controversial Cancer Star," *Science* 259 (1993), 1391–1392.

2. Ibid.

3. M. Chase, "As Genentech Awaits New Test of Old Drug, Its Pipeline Fills Up," *The Wall Street Journal*, 30 April 1993, A1, A4.

## Appendix
### The Human Genome Project

1. A. Kornberg, *For the Love of Enzymes* (Boston: Harvard University Press, 1989), 36.

2. J.D. Watson, "The Human Genome Project: Past, Present, and Future," *Science* 248 (1990), 44–51.

3. D.E. Koshland, "Sequences and Consequences of the Human Genome," *Science* 246 (1989), 189.

4. S.E. Luria, "Human Genome Program," *Science* 246 (1989), 873.

5. N. Angier, "Big Science, Is It

Worth the Price?" *The New York Times*, 5 June 1990, C5, C12.

6. R. Duncan and E.H. McConkey, "How Many Proteins Are There in a Typical Mammalian Cell?" *Clinical Chem.* 28 (1982), 749–755.

7. J.A. Egeland, D.S. Gerhard, D.L. Pauls, et al., "Bipolar Affective Disorders Linked to DNA Markers on Chromosome 11," *Nature* 325 (1987), 783–787.

8. Discussed in M. Barinaga, "Manic Depression Gene Put in Limbo," *Science* 246 (1989), 886–887.

# Credits

The quotes on page 3 are from page 132 of *Genes, Dreams, and Realities* by Sir Frank Macfarlane Burnet. Copyright © 1971 by Sir Macfarlane Burnet. Reprinted by permission of Basic Books, a division of HarperCollins, Publishers, Inc.; from information appearing in *The New England Journal of Medicine* 314 (1986), pp. 1226–1232, in "Progress Against Cancer?" by J.C. Bailar and E.M. Smith; and from a media advisory distributed by Fenton Communications on February 4, 1992, Washington, DC. Reprinted by permission of Dr. Sam Epstein.

The excerpt on page 6 is from an article by Dr. M.J. Wizenberg in *Cancer* 63 (1989), pp. 2387–2392. Reprinted by permission.

The quote on pages 6–7 is excerpted from information appearing in *The New England Journal of Medicine* 324 (1991), pp. 1180–1185, in "Survival and the Quality of Life Among Patients Receiving Unproven as Compared With Conventional Cancer Therapy," by B.R. Cassileth, E.J. Lusk, D. Guerry, et al.

The quote on page 11 is excerpted from information appearing in *The New England Journal of Medicine* 321 (1989), pp. 972–973, in "Training Basic Scientists to Bridge the Gap Between Basic Science and Its Application to Human Disease," by I.M. Arias.

The quote attributed to Pierre Chambon on page 23 is cited in "Three Europeans Find Their Own Road to Fame," by L. Balter, in *Science* 254 (1991), pp. 1116–1117. Copyright © 1991 by the American Association for the Advancement of Science.

The excerpt on page 34 is from "Human Cancer Research," by G.B. Dermer, in *Science* 221 (1983), p. 318. Copyright © 1983 by the American Association for the Advancement of Science.

# Index

# About the Author

Dr. Gerald B. Dermer is a research scientist, independent consultant, and writer specializing in cancer, pathology, and biotechnology. He received his bachelor's degree in biophysics, master's degree in genetics, and doctorate in cell biology from the University of California—Los Angeles, where he studied under the prominent electron microscopist, Dr. Fritiof S. Sjostrand. He then did postdoctoral work in Sweden under a fellowship awarded by the National Institutes of Health.

Dr. Dermer began his work on human cancer at the Hospital of the Good Samaritan in Los Angeles, where he had been hired to establish a laboratory of electron microscopy. He has spent over fifteen years involved in basic and clinical research on human tumors and published some forty papers on his findings in respected peer-reviewed medical and scientific journals, including *Cancer* and *Cancer Research*. In the course of his professional career, Dr. Dermer has also served as associate clinical professor of pathology at the University of Southern California School of Medicine and as associate professor of pathology at the North Carolina School of Medicine at Chapel Hill. He now lives in Mesa, Arizona.